INFORMATION TECHNOLOGY FOR COUNTERTERRORISM

IMMEDIATE ACTIONS AND **FUTURE POSSIBILITIES**

Committee on the Role of Information Technology
in Responding to Terrorism

Computer Science and Telecommunications Board

NATIONAL RESEARCH COUNCIL
OF THE NATIONAL ACADEMIES

John L. Hennessy, David A. Patterson, and Herbert S. Lin, Editors

THE NATIONAL ACADEMIES PRESS
Washington, D.C.
www.nap.edu

THE NATIONAL ACADEMIES PRESS • 500 Fifth Street, N.W. • Washington, DC 20001

NOTICE: This project was approved by the Governing Board of the National Research Council, whose members are drawn from the councils of the National Academy of Sciences, the National Academy of Engineering, and the Institute of Medicine. The members of the committee responsible for this final report were chosen for their special competences and with regard for appropriate balance.

The study from which this report is largely derived was supported by private funds from the National Academies. The additional work required to produce this report was supported by core funding from the Computer Science and Telecommunications Board (CSTB). Core support for CSTB in this period was provided by the Air Force Office of Scientific Research, Department of Energy, National Institute of Standards and Technology, National Library of Medicine, National Science Foundation, Office of Naval Research, and the Cisco, Intel, and Microsoft corporations. Sponsors enable but do not influence CSTB's work. Any opinions, findings, conclusions, or recommendations expressed in this publication are those of the authors and do not necessarily reflect the views of the organizations or agencies that provide support for CSTB.

International Standard Book Number 0-309-08736-8
Library of Congress Control Number: 2003101593

Copies of this report are available from the National Academies Press, 500 Fifth Street, N.W., Lockbox 285, Washington, DC 20055; (800) 624-6242 or (202) 334-3313 in the Washington metropolitan area. Internet, http://www.nap.edu.

Additional copies of this report are available in limited quantity from the Computer Science and Telecommunications Board, National Research Council, 500 Fifth Street, N.W., Washington, DC 20001. Call (202) 334-2605 or e-mail the CSTB at cstb@nas.edu.

Copyright 2003 by the National Academy of Sciences. All rights reserved.

Printed in the United States of America
First Printing, March 2003
Second Printing, Febuary 2004

Suggested citation: Computer Science and Telecommunications Board, *Information Technology for Counterterrorism: Immediate Actions and Future Possibilities*, The National Academies Press, Washington, D.C., 2003.

THE NATIONAL ACADEMIES
Advisers to the Nation on Science, Engineering, and Medicine

The **National Academy of Sciences** is a private, nonprofit, self-perpetuating society of distinguished scholars engaged in scientific and engineering research, dedicated to the furtherance of science and technology and to their use for the general welfare. Upon the authority of the charter granted to it by the Congress in 1863, the Academy has a mandate that requires it to advise the federal government on scientific and technical matters. Dr. Bruce M. Alberts is president of the National Academy of Sciences.

The **National Academy of Engineering** was established in 1964, under the charter of the National Academy of Sciences, as a parallel organization of outstanding engineers. It is autonomous in its administration and in the selection of its members, sharing with the National Academy of Sciences the responsibility for advising the federal government. The National Academy of Engineering also sponsors engineering programs aimed at meeting national needs, encourages education and research, and recognizes the superior achievements of engineers. Dr. Wm. A. Wulf is president of the National Academy of Engineering.

The **Institute of Medicine** was established in 1970 by the National Academy of Sciences to secure the services of eminent members of appropriate professions in the examination of policy matters pertaining to the health of the public. The Institute acts under the responsibility given to the National Academy of Sciences by its congressional charter to be an adviser to the federal government and, upon its own initiative, to identify issues of medical care, research, and education. Dr. Harvey V. Fineberg is president of the Institute of Medicine.

The **National Research Council** was organized by the National Academy of Sciences in 1916 to associate the broad community of science and technology with the Academy's purposes of furthering knowledge and advising the federal government. Functioning in accordance with general policies determined by the Academy, the Council has become the principal operating agency of both the National Academy of Sciences and the National Academy of Engineering in providing services to the government, the public, and the scientific and engineering communities. The Council is administered jointly by both Academies and the Institute of Medicine. Dr. Bruce M. Alberts and Dr. Wm. A. Wulf are chair and vice chair, respectively, of the National Research Council.

www.national-academies.org

COMMITTEE ON THE ROLE OF INFORMATION TECHNOLOGY IN RESPONDING TO TERRORISM

JOHN HENNESSY, Stanford University, *Chair*
DAVID A. PATTERSON, University of California at Berkeley, *Vice Chair*
STEVEN M. BELLOVIN, AT&T Laboratories
W. EARL BOEBERT, Sandia National Laboratories
DAVID BORTH, Motorola Labs
WILLIAM F. BRINKMAN, Lucent Technologies (retired)
JOHN M. CIOFFI, Stanford University
W. BRUCE CROFT, University of Massachusetts at Amherst
WILLIAM P. CROWELL, Cylink Inc.
JEFFREY M. JAFFE, Bell Laboratories, Lucent Technologies
BUTLER W. LAMPSON, Microsoft Corporation
EDWARD D. LAZOWSKA, University of Washington
DAVID LIDDLE, U.S. Venture Partners
TOM M. MITCHELL, Carnegie Mellon University
DONALD NORMAN, Northwestern University
JEANNETTE M. WING, Carnegie Mellon University

Staff

HERBERT S. LIN, Senior Scientist and Study Director
STEVEN WOO, Program Officer
DAVID DRAKE, Senior Project Assistant

COMPUTER SCIENCE AND TELECOMMUNICATIONS BOARD
2002-2003

DAVID D. CLARK, Massachusetts Institute of Technology, *Chair*
ERIC BENHAMOU, 3Com Corporation
DAVID BORTH, Motorola Labs
JOHN M. CIOFFI, Stanford University
ELAINE COHEN, University of Utah
W. BRUCE CROFT, University of Massachusetts at Amherst
THOMAS E. DARCIE, AT&T Labs Research
JOSEPH FARRELL, University of California at Berkeley
JOAN FEIGENBAUM, Yale University
HECTOR GARCIA-MOLINA, Stanford University
WENDY KELLOGG, IBM Thomas J. Watson Research Center
BUTLER W. LAMPSON, Microsoft Corporation
DAVID LIDDLE, U.S. Venture Partners
TOM M. MITCHELL, Carnegie Mellon University
DAVID A. PATTERSON, University of California at Berkeley
HENRY (HANK) PERRITT, Chicago-Kent College of Law
DANIEL PIKE, Classic Communications
ERIC SCHMIDT, Google Inc.
FRED SCHNEIDER, Cornell University
BURTON SMITH, Cray Inc.
LEE SPROULL, New York University
WILLIAM STEAD, Vanderbilt University
JEANNETTE M. WING, Carnegie Mellon University

Staff

MARJORY S. BLUMENTHAL, Executive Director
HERBERT S. LIN, Senior Scientist
ALAN S. INOUYE, Senior Program Officer
JON EISENBERG, Senior Program Officer
LYNETTE I. MILLETT, Program Officer
CYNTHIA A. PATTERSON, Program Officer
STEVEN WOO, Dissemination Officer
JANET BRISCOE, Administrative Officer
RENEE HAWKINS, Financial Associate
DAVID PADGHAM, Research Associate
KRISTEN BATCH, Research Associate
PHIL HILLIARD, Research Associate
MARGARET HUYNH, Senior Project Assistant

DAVID DRAKE, Senior Project Assistant
JANICE SABUDA, Senior Project Assistant
JENNIFER BISHOP, Senior Project Assistant
BRANDYE WILLIAMS, Staff Assistant

For more information on CSTB, see its Web site at <http://www.cstb.org>, write to CSTB, National Research Council, 500 Fifth Street, N.W., Washington, DC 20001, call at (202) 334-2605, or e-mail at cstb@nas.edu.

Preface

Immediately following the events of September 11, 2001, the National Academies (including the National Academy of Sciences, the National Academy of Engineering, the Institute of Medicine, and the National Research Council) offered its services to the nation to formulate a scientific and technological response to the challenges posed by emerging terrorist threats that would seek to inflict catastrophic damage on the nation's people, its infrastructure, or its economy. Specifically, it supported a project that culminated in a report entitled *Making the Nation Safer: The Role of Science and Technology in Countering Terrorism* (The National Academies Press, Washington, D.C.) that was released on June 25, 2002. That project, chaired by Lewis M. Branscomb and Richard D. Klausner, sought to identify current threats of catastrophic terrorism, understand the most likely vulnerabilities in the face of these threats, and identify highly leveraged opportunities for contributions from science and technology to counterterrorism in both the near term and the long term.

Taking the material on information technology contained in *Making the Nation Safer* as a point of departure, the Committee on the Role of Information Technology in Responding to Terrorism, identical to the Panel on Information Technology that advised the Branscomb-Klausner committee, drew on sources, resources, and analysis unavailable to that committee during the preparation of its report. In addition, the present report contains material and elaborations that the Branscomb-Klausner committee did not have time to develop fully for the parent report. Both reports are aimed at spurring research in the science and technology communities to counter and respond to terrorist acts such as those experienced on September 11.

In addition to presenting material on information technology (IT), *Making the Nation Safer* includes chapters on nuclear and radiological threats, human and agricultural health systems, toxic chemicals and explosive materials, energy systems, transportation systems, cities and fixed infrastructure, and the response of people to terrorism. The present report focuses on IT—its role as part of the national infrastructure, suggested areas of research (information and network security, IT for emergency response, and information fusion), and the people and organizational aspects that are critical to the acceptance and use of the proposed solutions. Note that policy is not a primary focus of this report, although policy issues are addressed as needed to provide context for the research programs outlined here.

Information Technology for Counterterrorism draws on many past reports and studies of the Computer Science and Telecommunications Board (CSTB). These CSTB reports include *Cybersecurity Today and Tomorrow: Pay Now or Pay Later; Computers at Risk: Safe Computing in the Information Age; Embedded, Everywhere: A Research Agenda for Networked Systems of Embedded Computers; Realizing the Potential of C4I: Fundamental Challenges; Information Technology Research for Crisis Management;* and *Computing and Communications in the Extreme,* among others. Furthermore, the report leverages current CSTB studies on geospatial information, authentication technologies, critical infrastructure protection and the law, and privacy.

The Committee on the Role of Information Technology in Responding to Terrorism included current and past CSTB members as well as other external experts. The 16 committee members (see the appendix for committee and staff biographies) are experts in computer, information, Internet, and network security; computer and systems architecture; computer systems innovation, including interactive systems; national security and intelligence; telecommunications, including wireline and wireless; data mining and information fusion and management; machine learning and artificial intelligence; automated reasoning tools; information-processing technologies; information retrieval; networked, distributed, and high-performance systems; software; and human factors. To meet its charge, the committee met several times over a 2-month period and conducted extensive e-mail dialogue to discuss the report text.

As was the parent report, this focused report was developed quickly, with the intent of informing key decision makers with respect to the role of information technology in the homeland security effort. The treatment of any of the subjects in this report is far from comprehensive or exhaustive—instead, the report highlights those subject aspects that the committee deems critical at this time. Accordingly, the report builds on, and cites heavily, prior CSTB reports that more substantially address the relevant issues.

The committee wishes to thank the CSTB staff (Herbert Lin as study director, Steven Woo for research support, and D.C. Drake for administrative support) for developing coherent drafts from scraps of e-mail and brief notes from committee meetings.

> John L. Hennessy, *Chair*
> David A. Patterson, *Vice Chair*
> Committee on the Role of Information Technology
>   in Responding to Terrorism

Acknowledgment of Reviewers

This report has been reviewed in draft form by individuals chosen for their diverse perspectives and technical expertise, in accordance with procedures approved by the National Research Council's (NRC's) Report Review Committee. The purpose of this independent review is to provide candid and critical comments that will assist the institution in making the published report as sound as possible and to ensure that the report meets institutional standards for objectivity, evidence, and responsiveness to the study charge. The review comments and draft manuscript remain confidential to protect the integrity of the deliberative process. We wish to thank the following individuals for their participation in the review of this report:

Edward Balkovich, The RAND Corporation,
Richard Baseil, The MITRE Corporation,
Jules A. Bellisio, Telcordia,
Tom Berson, Anagram Laboratories,
James Gray, Microsoft,
Daniel Huttenlocher, Cornell University,
Richard Kemmerer, University of California at Santa Barbara,
Keith Marill, New York University Bellevue Hospital Center,
William Press, Los Alamos National Laboratory,
Fred Schneider, Cornell University, and
Edward Wenk, University of Washington.

Although the reviewers listed above provided many constructive comments and suggestions, they were not asked to endorse the conclusions or recommendations, nor did they see the final draft of the report before its release. The review of this report was overseen by R. Stephen Berry of the University of Chicago. Appointed by the NRC's Report Review Committee, he was responsible for making certain that an independent examination of this report was carried out in accordance with institutional procedures and that all review comments were carefully considered. Responsibility for the final content of this report rests entirely with the Computer Science and Telecommunications Board and the National Research Council.

Contents

EXECUTIVE SUMMARY 1

1 BACKGROUND AND INTRODUCTION 10
   1.1 What Is Terrorism?, 10
   1.2 The Role of Information Technology in National Life and in Counterterrorism, 11
   1.3 The Information Technology Infrastructure and Associated Risks, 12

2 TYPES OF THREATS ASSOCIATED WITH INFORMATION TECHNOLOGY INFRASTRUCTURE 15
   2.1 Attack on IT as an Amplifier of a Physical Attack, 15
   2.2 Other Possibilities for Attack Involving IT, 16
      2.2.1 Attacks on the Internet, 16
      2.2.2 Attacks on the Public Switched Network, 18
      2.2.3 The Financial System, 20
      2.2.4 Embedded/Real-Time Computing, 20
      2.2.5 Control Systems in the National Critical Infrastructure, 21
      2.2.6 Dedicated Computing Facilities, 23
   2.3 Disproportionate Impacts, 23
   2.4 Threats in Perspective: Possibility, Likelihood, and Impact, 24

3 INVESTING IN INFORMATION TECHNOLOGY RESEARCH 28
   3.1 Information and Network Security, 31
      3.1.1 Authentication, 33
      3.1.2 Detection, 35
      3.1.3 Containment, 37
      3.1.4 Recovery, 40
      3.1.5 Cross-cutting Issues in Information and Network Security Research, 41
   3.2 Systems for Emergency Response, 46
      3.2.1 Intra- and Interoperability, 47
      3.2.2 Emergency Deployment of Communications Capacity, 55
      3.2.3 Security of Rapidly Deployed Ad Hoc Networks, 57
      3.2.4 Information-Management and Decision-Support Tools, 58
      3.2.5 Communications with the Public During an Emergency, 59
      3.2.6 Emergency Sensor Deployment, 60
      3.2.7 Precise Location Identification, 61
      3.2.8 Mapping the Physical Aspects of the Telecommunications Infrastructure, 62
      3.2.9 Characterizing the Functionality of Regional Networks for Emergency Responders, 62
   3.3 Information Fusion, 63
      3.3.1 Data Mining, 68
      3.3.2 Data Interoperability, 69
      3.3.3 Natural Language Technologies, 69
      3.3.4 Image and Video Processing, 70
      3.3.5 Evidence Combination, 70
      3.3.6 Interaction and Visualization, 71
   3.4 Privacy and Confidentiality, 71
   3.5 Other Important Technology Areas, 75
      3.5.1 Robotics, 75
      3.5.2 Sensors, 76
      3.5.3 Simulation and Modeling, 78
   3.6 People and Organizations, 80
      3.6.1 Principles of Human-Centered Design, 81
      3.6.2 Organizational Practices in IT-Enabled Companies and Agencies, 89
      3.6.3 Dealing with Organizational Resistance to Interagency Cooperation, 91

3.6.4 Principles into Practice, 93
3.6.5 Implications for Research, 95

4 WHAT CAN BE DONE NOW? 97

5 RATIONALIZING THE FUTURE RESEARCH AGENDA 106

APPENDIX: BIOGRAPHIES OF COMMITTEE AND
STAFF MEMBERS 115

WHAT IS CSTB? 127

Executive Summary

Making the Nation Safer: The Role of Science and Technology in Countering Terrorism, a report released by the National Academies in June 2002,[1] articulated the role of science and technology in countering terrorism. That report included material on the specific role of information technology (IT). Building on that report as a point of departure, the panel of experts responsible for the IT material in *Making the Nation Safer* was reconvened as the Committee on the Role of Information Technology in Responding to Terrorism in order to develop the present report.

DEFINING TERRORISM FOR THE PURPOSES OF THIS REPORT

Terrorism can occur on many different scales and with a wide range of impacts. While a terrorist act can involve a lone suicide bomber or a rental truck loaded with explosives, Americans' perception of catastrophic terrorist acts will forever be measured against the events of September 11, 2001. In one single day, thousands of lives and tens of billions of dollars were lost to terrorism. This report focuses primarily on the high-impact catastrophic dimensions of terrorism as framed by the events of September 11. Thus, in an IT context, the "lone hacker," or even the cybercriminal—while bothersome and capable of doing damage—is not the focus of this report. Instead, the report considers the larger threat posed

---

[1] National Research Council. 2002. *Making the Nation Safer: The Role of Science and Technology in Countering Terrorism.* The National Academies Press, Washington, D.C.

by smart, disciplined adversaries with ample resources. (Of course, measures taken to defend against catastrophic terrorism will likely have application in defending against less sophisticated attackers.)

THE ROLE OF INFORMATION TECHNOLOGY IN SOCIETY AND IN COUNTERTERRORISM

Information technology is essential to virtually all of the nation's critical infrastructures, from the air-traffic-control system to the aircraft themselves, from the electric-power grid to the financial and banking systems, and, obviously, from the Internet to communications systems. In sum, this reliance of all of the nation's critical infrastructures on IT makes any of them vulnerable to a terrorist attack on their computer or telecommunications systems.

An attack involving IT can take different forms. The IT itself can be the target. Or, a terrorist can either launch or exacerbate an attack by exploiting the IT infrastructure, or use IT to interfere with attempts to achieve a timely response. Thus, IT is both a target and a weapon. Likewise, IT also has a major role in counterterrorism—it can prevent, detect, and mitigate terrorist attacks. For example, advances in information fusion and data mining may facilitate the identification of important patterns of behavior that help to uncover terrorists or their plans in time to prevent attacks.

While there are many possible scenarios for an attack on some element(s) of the IT infrastructure (which includes the Internet, the telecommunications infrastructure, embedded/real-time computing such as SCADA [supervisory control and data acquisition] systems, and dedicated computing devices such as desktop computers), the committee believes that the most devastating consequences would occur if an attack on or using IT were part of a multipronged attack with other, more physical components. In this context, compromised IT could expand terrorist opportunities to widen the damage of a physical attack, diminish timely responses to the attack, and heighten terror in the population by providing false information about the nature of the threat.

The likelihood of a terrorist attack against or through the use of the IT infrastructure must be understood in the context of terrorists. Like other organizations, terrorist groups are likely to utilize their limited resources in activities that maximize impact and visibility. A decision by terrorists to use IT, or any other means, in an attack depends on factors such as the kinds of expertise and resources available, the publicity they wish to gain, and the symbolic value of an attack. How terrorists weigh such factors is not known in advance. Those wanting to create immediate public fear and terror are more likely to use a physical attack than an attack that targets IT exclusively.

WHAT CAN BE DONE NOW: SHORT-TERM RECOMMENDATIONS

The committee makes two short-term recommendations with respect to the nation's communications and information systems.

Short-Term Recommendation 1: The nation should develop a program that focuses on the communications and computing needs of emergency responders. Such a program would have two essential components:

- Ensuring that authoritative, current-knowledge expertise and support regarding IT are available to emergency-response agencies prior to and during emergencies, including terrorist attacks.
- Upgrading the capabilities of the command, control, communications, and intelligence (C3I) systems of emergency-response agencies through the use of existing technologies. Such upgrades might include transitioning from analog to digital systems and deploying a separate emergency-response communications network in the aftermath of a disaster.

Short-Term Recommendation 2: The nation should promote the use of best practices in information and network security in all relevant public agencies and private organizations.

- *For IT users on the operational level:* Ensure that adequate information-security tools are available. Conduct frequent, unannounced red-team penetration testing of deployed systems. Promptly fix problems and vulnerabilities that are known. Mandate the use of strong authentication mechanisms. Use defense-in-depth in addition to perimeter defense.
- *For IT vendors:* Develop tools to monitor systems automatically for consistency with defined secure configurations. Provide well-engineered schemes for user authentication based on hardware tokens. Conduct more rigorous testing of software and systems for security flaws.
- *For the federal government:* Position critical federal information systems as models for good security practices. Remedy the failure of the market to account adequately for information security so that appropriate market pro-security mechanisms develop.

WHAT CAN BE DONE IN THE FUTURE

Because the possible attacks on the nation's IT infrastructure vary so widely, it is difficult to argue that any one type is more likely than others. This fact suggests the value of a long-term commitment to a strategic research and development program that will increase the overall robustness of the computer and telecommunications networks. Such a program could improve the nation's ability to prevent, detect, respond to, and recover from terrorist attacks. This agenda would also have general applications, such as reducing cybercrime and responding to natural disasters. Three critical areas of research are information and network security, C3I systems for emergency response, and information fusion. Although technology is central to these three areas, it is not the sole element of concern. Research in these areas must be multidisciplinary, involving technologists, social scientists, and domain experts. Since technology deployed for operational purposes is subject to the reality of implementation and use by humans, technology cannot be studied in isolation from how it is deployed and used.

Information and Network Security

Research in information and network security is relevant to the nation's counterterrorism efforts for several reasons. IT attacks can amplify the impact of physical attacks and lessen the effectiveness of emergency responses. IT attacks on SCADA systems could be devastating. The increasing levels of social and economic damage caused by cybercrime suggest a corresponding increase in the likelihood of severe damage through cyberattacks. The technology discussed here is relevant to fighting cybercrime and to conducting efforts in defensive information warfare.

Research in information and network security can be grouped in four areas: authentication, detection, containment, and recovery; a fifth set of topics such as dealing with buggy code is broadly applicable.

- *Authentication* is relevant to better ways of preventing unauthorized parties from gaining access to a computer system to cause harm.
- *Detection* of intruders with harmful intentions is critical for thwarting their actions. However, because intruders take great care to hide their entry and/or make their behavior look innocuous, such detection is a very challenging problem (especially when the intruder is an insider gone bad).
- *Containment* is necessary if the success of an attacker is to be limited in scope. Although the principle of graceful degradation under attack is well accepted, system and network design for graceful degradation is not well understood.

EXECUTIVE SUMMARY

- *Recovery* involves backup and decontamination. In a security context, backup methods for use under adversarial conditions and applicable to large systems are needed. Decontamination—the process of distinguishing the clean system state from the infected portions and eliminating the causes of those differences—is especially challenging when a system cannot be shut down.
- *Other areas.* Buggy code (i.e., flawed computer programs) is probably the oldest unsolved problem in computer science, and there is no particular reason to think that research can solve the problem once and for all. One approach to the problem is to provide incentives to install fixes, even though the fixes themselves may carry risks such as exposing other software flaws. Many system vulnerabilities result from *improper administration*, and better system administration tools for specifying security policies and checking system configurations are necessary. Research in *tools for auditing functionality* to ensure that hardware and software have the prescribed—and no additional—functionality would be helpful. *Security that is more transparent* would have higher adoption rates. Understanding the *failure in the marketplace* of previous attempts to build in computer security would help guide future research efforts.

IT and C3I for Emergency Response

C3I systems are critical to emergency responders for coordinating their efforts and increasing the promptness and effectiveness of their response. C3I for emergency response to terrorist attacks poses challenges that differ from natural disasters: the number of responding agencies—from local, state, and federal governments—increases the degree of complexity, while the additional security or law-enforcement presence that is required may interfere with rescue and recovery operations.

C3I systems for emergency responders face many challenges:

- Regarding *ad hoc interoperability*, different emergency responders must be able to communicate with each other and other agencies, and poor interoperability among responding agencies is a well-known problem. Thus, for example, there is a technical need for protocols and technology that can facilitate interconnection and interoperation.
- Emergency situations result in extraordinary demands on *communications capacity*. Research is needed on using residual capacity more effectively and deploying additional ("surge") capacity.
- In responding to disasters, emergency-response managers need *decision-support tools* that can assist them in sorting, evaluating, filtering, and integrating information from a vast array of voice and data traffic.
- During an emergency, providing *geographically sensitive public*

information that is relevant to where people are (e.g., for evacuation purposes) is a challenging technical problem.

- *Sensors deployed in an emergency* could track the spread of nuclear or biological contaminants, locate survivors (e.g., through heat emanations or sounds), and find pathways through debris.
- *Location identification* of people and structures is a major problem when there is physical damage to a structure or an area.

Information Fusion

Information fusion promises to play a central role in the prevention, detection, and response to terrorism. For example, the effectiveness of checkpoints such as airline boarding gates could be improved significantly by creating information-fusion tools to support checkpoint operators in real time (a prevention task). Also, advances in the automatic interpretation of image, video, and other kinds of unstructured data could aid in detection. Finally, early response to biological attacks could be supported by collecting and analyzing real-time data such as admissions to hospital emergency rooms and purchases of nonprescription drugs in grocery stores. The ability to acquire, integrate, and interpret a range and volume of data will support decision makers such as emergency-response units and intelligence organizations.

Data mining is a technology for analyzing historical and current online data to support informed decision making by learning general patterns from a large volume of specific examples. But to be useful for counterterrorist purposes, such efforts must be possible over data in a variety of different and nonstructured formats, such as text, image, and video in multiple languages. In addition, new research is needed to normalize and combine data collected from multiple sources to improve *data interoperability*. And, new techniques for *data visualization* will be useful in exploiting human capabilities for pattern recognition.

Privacy and Confidentiality

Concerns over privacy and confidentiality are magnified in a counterterrorism intelligence context. The perspective of intelligence gatherers, "collect everything in case something might be useful," conflicts with the pro-privacy tenet of "don't collect anything unless you know you need it." To resolve this conflict, research is needed to provide policy makers with accurate information about the impact on privacy and confidentiality of different kinds of data disclosure. Furthermore, the development of new privacy-sensitive techniques may make it possible to provide useful information to analysts without compromising individual privacy. A variety of policy actions could also help to reduce the consequences of privacy violations.

Other Important Technology Areas

This report also briefly addresses three other technology areas: robotics, sensors, and modeling and simulation:

- *Robots,* which can be used in environments too dangerous for human beings, combine complex mechanical, perceptual, and computer and telecommunications systems, and pose significant research challenges such as the management of a team of robots and their integration.
- *Sensors,* used to detect danger in the environment, are most effective when they are linked in a distributed sensor network, a problem that continues to pose interesting research problems.
- *Modeling and simulation* can play important roles throughout crisis-management activities by making predictions about how events might unfold and by testing alternative operational choices. A key challenge is understanding the utility and limitations of models hastily created in response to an immediate crisis.

People and Organizations

Technology is always used in some social and organizational context, and human culpability is central in understanding how the system might succeed or fail. The technology cannot be examined in isolation from how it is deployed. Technology aimed at assisting people is essential to modern everyday life. At the same time, if improperly deployed, the technology can actually make the problem worse; human error can be extremely costly in time, money, and lives. Good design can dramatically reduce the incidence of error.

Principles of Human-Centered Design

Systems must be designed from a holistic, systems-oriented perspective. Principles that should guide such design include the following:

- *Put human beings "in the loop" on a regular basis.* Systems that use human beings only when automation is incapable of handling a situation are invariably prone to "human error."
- *Avoid common-mode failures, and recognize that common modes are not always easy to detect.*
- *Observe the distinction between work as prescribed and work as practiced.*

Procedures that address work as prescribed (e.g., tightening procedures and requiring redundant checking) often interfere with getting work done (i.e., work as practiced).

- *Probe security measures independently using tiger teams.* Tiger-team efforts, undertaken to test an organization's operational security posture using teams that simulate what a determined attacker might do, do what is necessary in order to penetrate security.

Organizational Resistance to Interagency Cooperation

An effective response to a serious terrorist incident will inevitably require interagency cooperation. However, because different agencies develop—and could reasonably be expected to develop—different internal cultures for handling the routine situations that they mostly address, interagency cooperation in a large-scale disaster is likely to be difficult under the best of circumstances.

There are no easy answers for bridging the cultural gulfs between agencies that are seldom called upon to interact. Effective interagency cooperation in times of crisis requires strong, sustained leadership that places a high priority on such cooperation and is willing to expend budget and personnel resources in support of it. Exercises and activities that promote interagency cooperation help to identify and solve some social, organizational, and technical problems, and also help to reveal the rivalries between agencies.

Research Implications Associated with Human and Organizational Factors

To better integrate the insights of social science into operational IT systems, research is relevant in at least four different areas:

- Formulating of system development methods that are more amenable to the incorporation of domain knowledge and social science expertise;
- Translating social science research findings into guidelines and methods that are readily applied by the technical community;
- Developing reliable security measures that do not interfere with work processes of legitimate employees; and
- Understanding the IT issues related to the disparate organizational cultures of agencies that will be fused under the Department of Homeland Security.

RATIONALIZING THE LONG-TERM RESEARCH AGENDA

The committee is silent on which government agency would best support the proposed research agenda. However, the research agenda should be characterized by the following:

- Support of multidisciplinary problem-oriented research that is useful both to civilian and to military users;
- A deep understanding and assessment of vulnerabilities;
- A substantial effort in research areas with a long time horizon for payoff, and tolerance of research directions that may not promise immediate applicability;
- Oversight by a board or other entity with sufficient stature to attract top talent to work in the field and to provide useful feedback; and
- Attention to the human resources needed to sustain the counter-terrorism IT research agenda.

One additional attribute of this R&D infrastructure would be desirable: the ability of researchers to learn from each other in a relatively free and open intellectual environment. Constraining the openness of that environment such as with classified research would have negative consequences for the research itself. Yet the free and open dissemination of information has potential costs, as terrorists may obtain information that they can use against us. The committee believes (or at least hopes) that there are other ways of reconciling the undeniable tension, and calls for some thought to be given to a solution to this dilemma.

1

Background and Introduction

1.1 WHAT IS TERRORISM?

Terrorism is usually defined in terms of non-state-sponsored attacks on civilians, perpetrated with the intent of spreading fear and intimidation. Terrorism can occur on many different scales and can cause a wide range of impacts. For many Americans, the events of September 11, 2001, changed dramatically their perceptions of what terrorism could entail. In the space of a few hours, thousands of American lives were lost, and property damage in the tens of billions of dollars occurred—an obviously high-impact event. However, as illustrated by the subsequent anthrax attacks, widespread disruption of key societal functions, loss of public confidence in the ability of governmental institutions to keep society safe, widespread loss of peace of mind, and/or pervasive injury to a society's way of life also count as manifestations of "high impact." It is on such high-impact, catastrophic dimensions of terrorism that the Committee on the Role of Information Technology in Responding to Terrorism decided to concentrate in order to keep the analytical focus of this report manageable.

The committee does not mean to suggest that only events of the magnitude of those on September 11 are worth considering. But the committee is primarily addressing events that would result in long-lasting and/ or major financial or life-safety impacts and that would generally require a coordinated response among multiple agencies, or are in many other respects very complicated to manage. Damaging and destructive though individual attacks are, the digital equivalent of a single car bomb with conventional explosives (e.g., a single hacker breaking into a nominally unsecured system that does not tunnel into other critical systems) is not the primary focus of this report.

In the context considered here, the adversary must be conceptualized as a very patient, smart, and disciplined opponent with many resources (money, personnel, time) at its disposal. Thus, in an information technology context, the "lone hacker" threat—often described in terms of maladjusted teenage males with too much time on their hands—is not the appropriate model. Protection against "ankle biters" and "script kiddies" who have the technical skills and understanding as well as the time needed to discover and exploit vulnerabilities is of course worth some effort, but it is important as well to consider seriously the larger threat that potentially more destructive adversaries pose.

1.2 THE ROLE OF INFORMATION TECHNOLOGY IN NATIONAL LIFE AND IN COUNTERTERRORISM

Information technology (IT) is essential to virtually all of the nation's critical infrastructures, which makes any of them vulnerable to a terrorist attack on the computer or telecommunications networks of those infrastructures. IT plays a critical role in managing and operating nuclear-power plants, dams, the electric-power grid, the air-traffic-control system, and financial institutions. Large and small companies rely on computers to manage payroll, track inventory and sales, and perform research and development. Every stage in the distribution of food and energy from producer to retail consumer relies on computers and networks. A more recent trend is the embedding of computing capability in all kinds of devices and environments, as well as the networking of embedded systems into larger systems.[1] And, most obviously, IT is the technological underpinning of the nation's communications systems, from the local loop of "plain old telephone service" to the high-speed backbone connections that support data traffic. These realities make the computer and communications systems of the nation a critical infrastructure in and of themselves, as well as major components of other kinds of critical infrastructure, such as energy or transportation systems.

In addition, while IT per se refers to computing and communications technologies, the hardware and software (i.e., the technological artifacts

---

[1]Computer Science and Telecommunications Board, National Research Council. 2001. *Embedded, Everywhere: A Research Agenda for Networked Systems of Embedded Computers.* National Academy Press, Washington, D.C. (Note that most Computer Science and Telecommunications Board reports contain many references to relevant literature and additional citations.)

of computers, routers, operating systems, browsers, fiber-optic lines, and so on) are part of a larger construct that involves people and organizations. The display on a computer system presents information for a person who has his or her own psychological and emotional attributes and who is usually part of an organization with its own culture and standard operating procedures. Thus, to understand how IT might fail or how the use of IT might not achieve the objectives desired, it is always necessary to consider the larger entity in which the IT is embedded.

IT also has a major role in the prevention, detection, and mitigation of terrorist attacks.[2] This report focuses on two critical applications. First, emergency response involves the agencies, often state and local, that are called upon to respond to terrorist incidents—firefighters, police, ambulance, and other emergency health care workers, and so on. These agencies are critically reliant on information technology to communicate, to coordinate, and to share information in a prompt, reliable, and intelligible fashion. Second, information awareness involves promoting a broad knowledge of critical information in the intelligence community to identify important patterns of behavior. Advances in information fusion, which is the aggregation of data from multiple sources for the purpose of discovering some insight, may be able to uncover terrorists or their plans in time to prevent attacks. In addition to prevention and detection, IT may also help rapidly and accurately identify the nature of an attack and aid in responding to it more effectively.

1.3 THE INFORMATION TECHNOLOGY INFRASTRUCTURE AND ASSOCIATED RISKS

The IT infrastructure can be conceptualized as having four major elements: the Internet, the conventional telecommunications infrastructure, embedded/real-time computing (e.g., avionics systems for aircraft control, supervisory control and data acquisition [SCADA] systems con-

---

[2]Computer Science and Telecommunications Board, National Research Council, 1996, *Computing and Communications in the Extreme: Research for Crisis Management and Other Applications*, National Academy Press, Washington, D.C.; Computer Science and Telecommunications Board, National Research Council, 1999, *Information Technology Research for Crisis Management*, National Academy Press, Washington, D.C. For purposes of the present report, prevention is relevant to the period of time significantly prior to an attack; during that period, a pending attack can be identified and the terrorist planning process for that attack disrupted or preempted. Detection is relevant in the period of time immediately before or during an attack (since an attack must first be detected before a response occurs). Mitigation is relevant during the time immediately after an attack, and it generally involves actions related to damage and loss minimization, recovery, and reconstitution.

trolling electrical energy distribution), and dedicated computing devices (e.g., desktop computers).

Each of these elements plays a different role in national life, and each has different specific vulnerabilities. Nevertheless, the ways in which IT can be damaged fall into three categories.[3] A system or network can become:

- *Unavailable.* That is, using the system or network at all becomes very difficult or impossible. The e-mail does not go through, or the computer simply freezes, or response time becomes intolerably long.
- *Corrupted.* That is, the system or network continues to operate, but under some circumstances of operation, it does not provide accurate results or information when one would normally expect. Alteration of data, for example, could have this effect.
- *Compromised.* That is, someone with bad intentions gains access to some or all of the capabilities of the system or network or the information available through it. The threat is that such a person could use privileged information or system control to further his or her malign purposes.

These types of damage are not independent—for example, an attacker could compromise a system in order to render it unavailable.

Different attackers might have different intentions with respect to IT. In some cases, an element of the IT infrastructure itself might be a target to be destroyed (e.g., the means for people to communicate or to engage in financial transactions). Alternatively, the target of the terrorist might be another kind of critical infrastructure (e.g., the electric-power grid), and the terrorist could either launch or exacerbate the attack by exploiting the IT infrastructure, or use it to interfere with attempts to achieve a timely and effective response.

In short, IT is both a target and a weapon that can be deployed against other targets. Counterterrorist activities thus seek to reduce the likelihood that IT functionality will be diminished as a result of an attack or as a result of the damage that might come from the use of IT as a weapon against valued targets.

A terrorist attack that involves the IT infrastructure can operate in one of several modes. First, an attack can come in "through the wires" as a hostile program (e.g., a virus or a Trojan horse program) or as a denial-

---

[3]Computer Science and Telecommunications Board, National Research Council. 2002. *Cybersecurity Today and Tomorrow: Pay Now or Pay Later.* National Academy Press, Washington, D.C.

of-service attack.[4] Second, some IT element may be physically destroyed (e.g., a critical data center or communications link blown up) or compromised (e.g., IT hardware surreptitiously modified in the distribution chain). Third, a trusted insider may be compromised (such a person, for instance, may provide passwords that permit outsiders to gain entry);[5] such insiders may also be conduits for hostile software or hardware modifications. All of these modes are possible and, because of the highly public and accessible nature of our IT infrastructure and of our society in general, it is impossible to fully secure this infrastructure against them. Nor are they mutually exclusive, and in practice they can be combined to produce even more destructive effects.

---

[4] A "through-the-wires" attack is conducted entirely at a distance and requires no physical proximity to the target.

[5] Computer Science and Telecommunications Board, National Research Council. 1999. *Trust in Cyberspace*. National Academy Press, Washington, D.C.

2

Types of Threats Associated with Information Technology Infrastructure

Most of the nation's civil communications and data network infrastructure is not hardened against attack, but this infrastructure tends to be localized either in geography or in mode of communication. Thus, if no physical damage is done to them, the computing and communications capabilities disrupted in an attack are likely to be recoverable in a relatively short time. Although their scope or scale is limited, they are nonetheless potentially attractive targets for what might be called "incremental" terrorism. That is, terrorists could use IT as the weapon in a series of relatively local attacks that are repeated against different targets—such as banks, hospitals, or local government services—so often that public confidence is shaken and significant economic disruption results.

However, this report focuses primarily on catastrophic terrorism, and the committee's analysis is aimed at identifying threats of that magnitude in particular and at proposing science and technology (S&T) strategies for combating them. Of course, serious efforts are needed to develop and deploy security technologies to harden all elements of the IT infrastructure to reduce the potential for damage from repeated attacks.

2.1 ATTACK ON IT AS AN AMPLIFIER OF A PHYSICAL ATTACK

Given IT's critical role in many other elements of the national infrastructure and in responding to crises, the committee believes that the targeting of IT as part of a multipronged attack scenario could have the most catastrophic consequences. Compromised IT can have several disastrous effects: expansion of terrorists' opportunities to widen the damage of a physical attack (e.g., by providing false information that drives people toward, rather than away from, the point of attack); diminution of timely responses to an attack (e.g., by interfering with communications systems of first responders); and heightened terror in the population through misinformation (e.g., by providing false information about the nature of a threat). The techniques to compromise key IT systems—for example, launching distributed denial-of-service (DDOS) attacks against Web sites and servers of key government agencies at the federal, state, and local levels; using DDOS attacks to disrupt agencies' telephone services and the emergency-response 911 system; or sending e-mails containing false information with forged return addresses so that they appear to be from trusted sources—are fairly straightforward and widely known.

2.2 OTHER POSSIBILITIES FOR ATTACK INVOLVING IT

When an element of the IT infrastructure is directly targeted, the goal is to destroy a sufficient amount of IT-based capability to have a significant impact, and the longer that impact persists, the more successful it is from the terrorist's point of view. For example, one might imagine attacks on the computers and data storage devices associated with important facilities. Irrecoverable loss of critical operating data and essential records on a large scale would likely result in catastrophic and irreversible damage to the U.S. economy. However, most major businesses already have disaster-recovery plans in place that include the backup of their data in a variety of distributed and well-protected locations (and in many cases, they augment backups of data with backup computing and communications facilities).[1] While no law of physics prevents the simultaneous destruction of all data backups and backup facilities in all locations, such an attack would be highly complex and difficult to execute and is thus highly unlikely.

2.2.1 Attacks on the Internet

The infrastructure of the Internet is another possible terrorist target, and given the Internet's public prominence, it may appeal to terrorists as an attractive target. The Internet could be seriously degraded for a relatively short period of time by a denial-of-service attack,[2] but such impact

---

[1]On the other hand, backup sites are often shared—one site may protect the data of multiple firms.

[2]A denial-of-service attack floods a target with a huge number of requests for service, thus keeping it busy servicing these (bogus) requests and unable to service legitimate ones.

is unlikely to be long lasting. The Internet itself is a densely connected network of networks that automatically routes around links that become unavailable,[3] which means that a large number of important nodes would have to be destroyed simultaneously to bring it down for an extended period of time. Destruction of some key Internet nodes could result in reduced network capacity and slow traffic across the Internet, but the ease with which Internet communications can be rerouted would minimize the long-term damage.[4] (In this regard, the fact that substantial data-networking services survived the September 11 disaster despite the destruction of large amounts of equipment—concentrated in the World Trade Center complex—reflected redundancies in the infrastructure and a measure of good fortune as well.)

The terrorist might obtain higher leverage with a "through-the-wires" attack that would require the physical replacement of components in Internet relay points on a large scale,[5] though such attacks would be much harder to plan and execute. Another attack that would provide higher leverage is on the Internet's Domain Name System (DNS), which translates domain names (e.g., example.com) to specific Internet Protocol (IP) addresses (e.g., 192.0.34.72) denoting specific Internet nodes. A relatively small number of "root name servers" underpins the DNS. Although the DNS is designed to provide redundancy in case of accidental failure, it has some vulnerability to an intentional physical attack that might target all name servers simultaneously. Although Internet operations would not halt instantly, an increasing number of sites would, over a period of time measured in hours to days, become inaccessible without root name servers to provide authoritative translation information. However, recovery from such an attack would be unlikely to take more than several days—damaged servers can be replaced, since they are general-purpose computers that are in common use.

In addition, most companies today do not rely on the Internet to carry out their core business functions. Even if a long-term disruption to the Internet were a major disruption to an e-commerce company such as Amazon.com or Dell, most other companies could resort to using phones

---

[3]Computer Science and Telecommunications Board, National Research Council. 2001. *The Internet's Coming of Age.* National Academy Press, Washington, D.C. Note, however, that the amount of redundancy is primarily limited by economic factors.

[4]This comment largely applies to U.S. use of the Internet. It is entirely possible that other nations—whose traffic is often physically routed through one or two locations in the United States—would fare much worse in this scenario.

[5]For example, many modern computers allow certain hardware components to be reprogrammed under software control. Improper use of this capability can damage hardware permanently.

and faxes again to replace the Internet for many important functions. (For example, the Department of the Interior has been largely off the Internet since December 5, 2001,[6] but it has continued to operate more or less as usual.)

Because the Internet is not yet central to most of American society, the impact of even severe damage to the Internet is less than what might be possible through other modes of terrorist attack. However, current trends suggest that the reliance on the Internet for key functions is likely to grow in the future, despite the existence of real security threats, and so this assessment about lower levels of impact from attacks on the Internet may become less valid in the future.

Box 2.1 provides some historical examples of attacks on the Internet.

2.2.2 Attacks on the Public Switched Network

The telecommunications infrastructure of the public switched network is likely to be less robust than the Internet. Although the long-haul telecommunications infrastructure is capable of dealing with single-point failures (and perhaps even double-point failures) in major switching centers, the physical redundancy in that infrastructure is finite, and damaging a relatively small number of major switching centers for long-distance telecommunications could result in a fracturing of the United States into disconnected regions.[7] Particular localities may be disrupted for a considerable length of time—in the aftermath of the September 11 attacks in New York City, telephone service in the downtown area took months to restore fully. Note also that many supposedly independent circuits are trenched together in the physical trenches along certain highway and rail rights-of-way, and thus these conduits constitute not just "choke points" but rather "choke routes" that are hundreds of miles long and that could be attacked anywhere.

An additional vulnerability in the telecommunications infrastructure is the local loop connecting central switching offices to end users; full recovery from the destruction of a central office entails the tedious rewiring of tens or hundreds of thousands of individual connections. Destruction of central offices on a large scale is difficult, simply because even an individual city has many of them, but destruction of a few central offices

---

[6]For additional information, see <http://www.computerworld.com/storyba/0,4125,NAV47_STO66665,00.html>.

[7]An exacerbating factor is that many organizations rely on leased lines to provide high(er)-assurance connectivity. However, these lines are typically leased from providers of telecommunications infrastructure and hence suffer from many of the same kinds of vulnerabilities as those that affect ordinary lines.

BOX 2.1   Historical Examples of Attacks on the Internet

- In March 1999, the Melissa virus infected e-mail systems worldwide. According to estimates from federal officials, the virus caused $80 million in disruption, lost commerce, and computer downtime, and infected 1.2 million computers. The virus launched when a user opened an infected Microsoft Word 8 or Word 9 document contained in either Office 97 or Office 2000.[1] The virus, programmed as a macro in the Word document, prompted the Outlook e-mail program to send the infected document to the first 50 addresses in the victim's Outlook address book. When a recipient opened the attachment in the e-mail, which appeared to be from a friend, co-worker, boss or family member, the virus spread to the first 50 e-mail addresses in that person's address book, and thus continued to propagate. Six months after the first appearance of Melissa, variant strains continued to make their way into users' inboxes despite warnings and widespread publicity about opening attachments while the macro function is enabled.

- Over a four-day period beginning February 7, 2000, distributed denial-of-service attacks temporarily shut down Yahoo, Amazon, E*Trade, eBay, CNN.com, and other Web sites. Yahoo shut down its site for several hours during peak viewing hours at an estimated cost of $116,000.[2] While the companies behind the targeted Web sites said that the attacks themselves would have minor financial impact, the attacks were of such importance that the White House convened a group of computer-security experts and technology executives to discuss the Internet's vulnerabilities. Federal officials spent millions in investigations of the DDOS attacks that garnered significant public attention.

- On July 19, 2001, the Code Red program "worm" infected more than 359,000 computers in less than 14 hours and 2,000 new infections per minute occurred during the height of its attack. Nimda, a similar hostile program first appearing on September 18, 2001, was potentially more damaging because it combined successful features of previous viruses such as Melissa and ILOVEYOU. During the first 24 hours, Nimda spread through e-mail, corporate networks, and Web browsers, infecting as many as 150,000 Web server and personal computers (PCs) in the United States. The virus—"admin" spelled backwards—was designed to affect PCs and servers running the Windows operating system and to resend itself every 10 days unless it was deleted. Nimda reproduced itself both via e-mail and over the Web—a user could be victimized by merely browsing a Web site that was infected. Furthermore, the infected machines sent out a steady stream of probes looking for new systems to attack. The additional traffic could effectively shut down company networks and Web sites; Nimda-generated traffic did not slow down the Internet overall, but infected companies reported serious internal slowdowns.[3]   Code Red and Nimda are examples of these new blended threats. Both are estimated to have caused $3 billion worldwide in lost productivity and for testing, cleaning, and deploying patches to computer systems.[4]

---

[1] Ann Harrison. 1999. "FAQ: The Melissa Virus," COMPUTERWORLD, March 31. Available online at <http://www.computerworld.com/news/1999/story/0,11280,27617,00.html>.

[2] Ross Kerber. 2000. "Vandal Arrests Would Only Be the Beginning Penalties, Damages Seen Hard to Determine," The Boston Globe, February 11.

[3] Henry Norr. 2001. "New Worm Plagues Systems Worldwide," The San Francisco Chronicle, September 19.

[4] Gregory Hulme. 2002. "One Step Ahead—Security Managers Are Trying to Be Prepared for the Next Blended Threat Attack," InformationWeek, May 20.

associated with key facilities or agencies (e.g., those of emergency-response agencies or of the financial district) would certainly have a significant immediate though localized impact. However, the widespread availability of cellular communications, and mobile base-stations deployable in emergency conditions, may mitigate the effect of central office losses.

2.2.3 The Financial System

The IT systems and networks supporting the nation's financial system are undeniably critical. The financial system is based on the Federal Reserve banking system, a system for handling large-value financial transactions (including Fedwire operated by the Federal Reserve, CHIPS, and SWIFT), and a second system for handling small-value retail transactions (including the Automated Clearing House, the credit-card system, and paper checks).[8] By its nature, the system for retail transactions is highly decentralized, while the system for large-value transactions is more centralized. Both the Federal Reserve system and the system for large-value transactions operate on networks that are logically distinct from the public telecommunications system or the Internet, and successful information attacks on these systems likely necessitate significant insider access.[9]

2.2.4 Embedded/Real-Time Computing

Embedded/real-time computing in specific systems could be attacked. For example, many embedded computing systems could be corrupted over time.[10] Of particular concern could be avionics in airplanes,

---

[8] For an extended (though dated) discussion of the infrastructure underlying the financial system, see John C. Knight et al., 1997, *Summaries of Three Critical Infrastructure Applications*, Computer Science Report No. CS-97-27, Department of Computer Science, University of Virginia, Charlottesville, November 14.

[9] The fact that these networks are logically separate from those of the Internet and the public switched telecommunications network reduces the risk of penetration considerably. In addition, security consciousness is much higher in financial networks than it is on the Internet. On the other hand, the fact that these networks are much smaller than the Internet suggests that there is less redundancy in them and that the computing platforms are likely to be less diverse compared with those on the Internet, a factor that tends to reduce security characteristics as compared with those of the Internet. Also, the physical infrastructure over which these financial networks communicate is largely shared, which means that they are vulnerable to large-scale physical disruptions or attacks on the telecommunications infrastructure.

[10] An inadvertent demonstration of this possibility was illustrated with the Y2K problem that was overlooked in many embedded/real-time systems designed in the 1980s and earlier.

collision-avoidance systems in automobiles, and other transportation systems. Such attacks would require a significant insider presence in technically responsible positions in key sectors of the economy over long periods of time. Another example is that sensors, which can be important elements of counterterrorism precautions, could be the target of an attack or, more likely, precursor targets of a terrorist attack.

2.2.5 Control Systems in the National Critical Infrastructure

Another possible attack on embedded/real-time computing would be an attack on the systems controlling elements of the nation's critical infrastructure, for example, the electric-power grid, the air-traffic-control system, the financial network, and water purification and delivery. An attack on these systems could trigger an event, and conceivably stimulate an inappropriate response that would drive large parts of the the overall system into a catastrophic state. Still another possibility is the compromise or destruction of systems and networks that control and manage elements of the nation's transportation infrastructure; such an attack could introduce chaos and disruption on a large scale that could drastically reduce the capability of transporting people and/or freight (including food and fuel).

To illustrate, consider the electric-power grid, which is one of the few, if not the only, truly *national* infrastructures in which it is theoretically possible that a failure in a region could cascade to catastrophic proportions before it could be dealt with. The electric-power grid is controlled by a variety of IT-based SCADA systems. (Box 2.2 describes some of the security issues associated with these systems.) Attacks on SCADA systems could obviously result in disruption of the network ("soft" damage), but because SCADA is used to control physical elements, such attacks could also result in irreversible physical damage. In cases in which backups for damaged components were not readily available (and might have to be remanufactured from scratch), such damage could have long-lasting impact. (Similar considerations apply to other parts of the nation's infrastructure.)

An electronic attack on a portion of the electric-power grid could result in significant damage, easily comparable to that associated with a local blackout. However, if terrorists took advantage of the chaos caused by a local blackout, they could likely inflict greater physical damage than would be possible in the absence of a blackout.

Another plausible disaster scenario that could rise to the level of catastrophic damage would be an attack on a local or regional power system that cascaded to shut down electrical power over a much wider area and possibly caused physical damage that could take weeks to repair.

BOX 2.2 Security Vulnerabilities and Problems of SCADA Systems

Today's supervisory control and data acquisition (SCADA) systems have been designed with little or no attention to security. For example, data in SCADA systems are often sent "in the clear." Protocols for accepting commands are open, with no authentication required. Control channels are often wireless or leased lines that pass through commercial telecommunications facilities. Unencrypted radio-frequency command pathways to SCADA systems are common and, for economic reasons, the Internet itself is increasingly used as a primary command pathway. In general, there is minimal protection against the forgery of control messages or of data and status messages. Such control paths present obvious vulnerabilities.

In addition, today's SCADA systems are built from commercial off-the-shelf components and are based on operating systems that are known to be insecure. Deregulation has meant placing a premium on the efficient use of existing capacity, and hence interconnections to shift supply from one location to another have increased. Problems of such distributed real-time dynamic control, in combination with the complex, highly interactive nature of the system being controlled, have become major issues in operating the power grid reliably.

A final problem arises because of the real-time nature of SCADA systems, in which timing may be critical to performance and optimal efficiency (timing is important because interrupts and other operations can demand millisecond accuracy): security add-ons in such an environment can complicate timing estimates and cause severe degradation to SCADA performance.

Compounding the difficulty of securing SCADA systems is the fact that information about their vulnerability is so readily available. Such information was first brought into general view in 1998-1999, when numerous details on potential Y2K problems were put up on the World Wide Web. Additional information of greater detail—dealing with potential attacks that were directly or indirectly connected to the President's Commission on Critical Infrastructure Protection—was subsequently posted on Web pages as well. Product data and educational videotapes from engineering associations can be used to familiarize potential attackers with the basics of the grid and with specific elements. Information obtained through semiautomated reconnaissance to probe and scan the networks of a variety of power suppliers could provide terrorists with detailed information about the internal workings of the SCADA network, down to the level of specific makes and models of equipment used and version releases of corresponding software. And more inside information could be obtained from sympathetic engineers and operators.

By comparison with the possibility of an attack on only a portion of the power grid, the actual feasibility of an attack that would result in a cascading failure with a high degree of confidence is not clear; a detailed study both of SCADA systems and the electric-power system would probably be required in order to assess this possibility. However, because of the inordinate complexity of the nation's electric-power grid, it would be difficult for either grid operators or terrorists to predict with any confidence the effects on the overall grid from a major disruptive event in one part of the system. Thus, any nonlocalized impact on the power grid would be as much a matter of chance as a foreseeable consequence.

2.2.6 Dedicated Computing Facilities

In many of the same ways that embedded computing could be attacked, dedicated computers such as desktop computers could also be corrupted in ways that are hard to detect. One possible channel comes from the extensive use of untrustworthy IT talent among software vendors.[11] Once working on the inside, perhaps after a period of years in which they act to gain responsibility and trust, it could happen that these individuals would be able to introduce additional but unauthorized functionality into systems that are widely used. Under such circumstances, the target might not be the general-purpose computer used in the majority of offices around the country, but rather the installation of hidden rogue code in particular sensitive offices. Another possible channel for attacking dedicated computing facilities results from the connection of computers through the Internet; such connections provide a potential route through which terrorists might attack computer systems that do provide important functionality for many sectors of the economy. Examples of widely used Internet-based vectors that, if compromised would have a large-scale effect in a short time, include the operating systems upgrades and certain shareware programs, such as those for sharing music files. (It is likely that Internet-connected computer systems that provide critical functionality to companies and organizations are better protected through firewalls and other security measures than is the average system on the Internet, but as press reports in recent years make clear, such measures do not guarantee that outsiders cannot penetrate them.)

2.3 DISPROPORTIONATE IMPACTS

Some disaster scenarios could result in significant loss or damage that is out of proportion to the actual functionality or capability destroyed. In

---

[11]Untrustworthy talent may be foreign or domestic in origin. Because foreign IT workers—whether working in the United States (e.g., under an H1-B visa or a green card) or offshore on outsourced work—are generally not subject to thorough background investigations, an obvious route is available through which foreign terrorist organizations can gain insider access. On the other hand, reports of American citizens having been successfully recruited by foreign terrorist organizations add a degree of believability to the scenario of domestic IT talent's being used to compromise systems for terrorist purposes.

particular, localized damage that resulted in a massive loss of confidence in some critical part of the infrastructure could have such a disproportionate impact. For example, if terrorists were able to make a credible claim that the control software of a popular "fly-by-wire" airliner was corrupted and could be induced to cause crashes on demand, perhaps demonstrating it once, public confidence in the airline industry might well be undermined. A more extreme scenario might be that the airlines themselves would ground airplanes until they could be inspected and the software validated.

To the extent that critical industries or sectors rely upon any element of the IT infrastructure, such disproportionate-impact disaster scenarios are a possibility. For this reason, certain types of attack that do not cause extensive actual damage must be considered to have some catastrophic potential. Accordingly, response plans must take into account how to communicate with the public for purposes of reassurance. (This point is beyond the scope of this report but is addressed in *Making the Nation Safer*.[12])

2.4 THREATS IN PERSPECTIVE: POSSIBILITY, LIKELIHOOD, AND IMPACT

While the scenarios described above are necessarily speculative, it is possible to make some judgments that relate to their likelihood:

- For a variety of reasons, state support of terrorism poses threats of a different and higher order of magnitude than does cybercrime or terrorism without state sponsorship. These reasons include access to large amounts of financial backing and the ability to maintain an actively adversarial stance at a high level for extended periods of time. For example, terrorists with the support of a state might be able to use the state's intelligence services to gain access to bribable or politically sympathetic individuals in key decision-making places or to systematically corrupt production or distribution of hardware or software.
- The most plausible threats are simple attacks launched against complex targets. The successful execution of complex attacks requires that many things go right, so simplicity in attack planning is an important consideration. Complex rather than simple targets are desirable because of the likelihood that the failure modes of a complex target are usually not

---

[12]National Research Council. 2002. *Making the Nation Safer: The Role of Science and Technology in Countering Terrorism*. National Academies Press, Washington, D.C.

well understood by its designers, and thus there are many more ways in which failure can occur in such systems.

- Attacks that require insider access are more difficult to carry out and thus less likely to occur than attacks that do not. Insiders must be placed or recruited and are not necessarily entirely trustworthy even from the standpoint of the attacker. Individuals with specialized expertise chosen to be placed as infiltrators may not survive the screening process, and because there is a limited number of such individuals, it can be difficult to insert an infiltrator into a target organization. In addition, compared to approaches not relying on insiders, insiders may leave behind more tracks that can call attention to their activities. This judgment depends, of course, on the presumed diligence of employers in ensuring that their key personnel are trustworthy, and it is worth remembering that the most devastating espionage episodes in recent U.S. history have involved insiders (i.e., Aldrich Ames and Robert Hanssen).

- Attacks that require execution over long periods of time are harder and thus less likely to mount than attacks that do not. Planning often takes place over a long period of time, but the actual execution of a plan can be long as well as short. When a plan requires extended activity that, if detected, would be regarded as abnormal, it is more likely to be discovered and/or thwarted.

- Terrorist attacks can be sustained over time as well as occurring in individual instances. If the effects of an attack sustained over time (perhaps over months or years) are cumulative, and if the attack goes undetected, the cumulative effects could reach very dangerous proportions. Because such an attack proceeds a little bit at a time, the resources needed to carry it out may well be less than those needed in more concentrated attacks, thus making it more feasible.

- Plans that call for repeated attacks are less likely to succeed than those calling for a single attack. For example, it is true that repeated attacks against the Internet could have effects that would defeat efforts to repair or secure it after one initial attack. Such an onslaught would be difficult to sustain, however, because it is highly likely to be detected, and efforts would be made to counter it. Instead, an adversary with the wherewithal to conduct such repeated attacks would more likely make the initial strike and then use the recovery period not to stage and launch another strike against the Internet but to attack the physical infrastructure; this strategy could leverage the inoperative Internet to cause additional damage and chaos. (Of course, the fact that physical attacks may be more difficult to conduct must also be taken into account.)

- The IT infrastructure (or some element of it) can be a weapon used in an attack on something else as well as being the target of an attack. An attack using the IT infrastructure as a weapon has advantages and disadvantages from the point of view of a terrorist planner. It can be conducted at a distance in relative physical safety, in a relatively anonymous fashion, and in potentially undetectable ways. However, the impact of such an attack (by assumption, on some other critical national asset) would be indirect, harder to predict, and less certain.

- Some of the scenarios described above are potentially relevant to information warfare attacks against the United States—that is, attacks launched or abetted by hostile nation-states and/or directed against U.S. military forces or assets. A hostile nation conducting an information attack on the United States is likely to conceal its identity to minimize the likelihood of retaliation, and thus it may resort to sponsoring terrorists who can attack without leaving clear national signatures.

The committee wishes to underscore a very important point regarding terrorist threats to the IT infrastructure—they are serious enough to warrant considerable national attention, but they are, in the end, only one of a number of ways through which terrorists could act against the United States. Thus, the likelihood of some kind of terrorist attack against or using the IT infrastructure must be understood in the context of a terrorist organization that may have many other types of attack at its disposal, including (possibly) chemical, biological, nuclear, radiological, suicide, and explosive attacks. This point is important because terrorists, like other parties, have limited resources. Thus, they are likely to concentrate their efforts where the impact is largest for the smallest expenditure of resources.

Many factors would play into a terrorist decision to use one kind of attack or another. The particular kinds of expertise and level of resources available, the effect that the terrorists wished to produce, the publicity they wished to gain, the complexity of any attack contemplated, the symbolic value of an attack, the risk of being caught, the likelihood of survival, the defenses that would be faced if a given attack was mounted, and the international reaction to such an attack are all relevant to such a decision. How any given terrorist will weigh such factors cannot be known in advance.

For example, terrorists who want to create immediate public fear and terror are more likely to use a physical attack (perhaps in conjunction with an attack using IT to amplify the resulting damage) than an attack that targets IT exclusively. The reason is that the latter is not likely to be as cinematic as other attacks. What would television broadcast? There would be no dead or injured people, no buildings on fire, no panic in the streets, and no emergency-response crews coming to the rescue. (This is not to say that an attack targeting IT exclusively could not shake public confidence—but it would not have the same impact as images of death and destruction in the streets.)

Note also that "likelihood" is not a static quantity. While it is true, all else being equal, that it is appropriate to devote resources preferentially to defending against highly likely attacks, the deployment of a defense that addresses the threat of a highly likely Attack A may well lead to a subsequent increase in the likelihood of a previously less likely Attack B. In short, terrorists may not behave in accordance with expectations that are based on static probability distributions. It is therefore very difficult to prioritize a research program for countering terrorism in the same way that one might, for example, prioritize a program for dealing with natural disasters.

How likely are terrorist attacks on the IT infrastructure or attacks using the IT infrastructure compared to terrorist attacks spreading smallpox or smuggling a stolen nuclear weapon into the United States? For obvious reasons, the committee is not in a position to make such judgments. But while the considerations discussed in this section make certain types of attack more or less likely, none of the scenarios described in Section 2.2 can be categorically excluded.

This fact argues in favor of a long-term commitment to a strategic R&D program that will contribute to the overall robustness of the telecommunications and data networks and of the platforms associated with them. Such a program would involve both fundamental research into the scientific underpinnings of information and network security as well as the development of deployable technology that would contribute to information and network security. Ultimately, the strengthening of the nation's IT infrastructure can improve our ability to prevent, detect, respond to, and recover from terrorist attacks on the nation.[13]

---

[13]Computer Science and Telecommunications Board, National Research Council, 1996, *Computing and Communications in the Extreme: Research for Crisis Management and Other Applications*, National Academy Press, Washington, D.C; Computer Science and Telecommunications Board, National Research Council, 1999, *Information Technology Research for Crisis Management*, National Academy Press, Washington, D.C.

3

Investing in Information Technology Research

This chapter describes the shape of a strategic research and development (R&D) program with respect to information technology for counterterrorism. However, it should be noted that this program has broad applicability not only for efforts against terrorism and information warfare but also for reducing cybercrime and responding to natural disasters. While the scope and complexity of issues with respect to each of these areas may well vary (e.g., a program focused on cybercrime may place more emphasis on forensics useful in prosecution), the committee believes that there is enough overlap in the research problems and approaches to make it unwise to articulate a separate R&D program for each area.

Although many areas of information technology research could be potentially valuable for counterterrorist purposes, the three areas described below are particularly important for helping reduce the likelihood or impact of a terrorist attack:

1. *Information and network security.* Research in information and network security is critically relevant to the nation's counterterrorism efforts for several reasons.[1] First, IT attacks can amplify the impact of physical

---

[1]Computer Science and Telecommunications Board (CSTB), National Research Council (NRC), 1991, *Computers at Risk: Safe Computing in the Information Age,* National Academy Press, Washington, D.C. (hereafter cited as CSTB, NRC, 1991, *Computers at Risk*); Computer Science and Telecommunications Board, National Research Council, 1999, *Trust in Cyberspace,* National Academy Press, Washington, D.C. (hereafter cited as CSTB, NRC, 1999,

attacks and lessen the effectiveness of emergency responses; reducing such vulnerabilities will require major advances in information and network security. IT attacks on supervisory control and data acquisition (SCADA) systems in the control infrastructure could also be highly damaging, and research to improve the security of such systems will be needed. Second, the increasing levels of social and economic damage caused by cybercrime and the tendency to rely on the Internet as the primary networking and communications channel both suggest that the likelihood of severe damage through a cyberattack is increasing. Finally, the evolution of the Internet and the systems connected to it demonstrates increasing homogeneity in hardware and software (Box 3.1), which makes the Internet more vulnerable at the same time that it becomes more critical. To address these problems, more researchers and trained professionals who are focused on information and network security will be needed.

2. *Systems for emergency response.* "C3I" (command, control, communications, and intelligence) systems are critical to emergency responders for coordinating efforts and increasing the promptness and effectiveness of response (e.g., saving lives, treating the injured, and protecting property). While terrorist attacks and natural disasters have many similarities with respect to the consequences of such events, the issues raised by C3I for emergency response for terrorist disasters differ from those for natural disasters for several reasons. First, the number of responding agencies, including those from the local, regional, state, and federal levels—with possibly conflicting and overlapping areas of responsibility—increases the level of complexity. (For example, in a terrorist attack scenario, the Department of Defense [DOD] might be much more involved than it would be in a natural disaster.) Second, ongoing security and law-enforcement concerns are much stronger in the wake of a terrorist attack. While looting is often a threat to the community affected by a natural disaster (and may result in the deployment of a police presence in the midst of the recovery effort), the threat from a follow-on terrorist attack may well be much greater or more technologically sophisticated than that posed by looters. And, to the extent that an additional security or law-enforcement presence is required, the sometimes-conflicting needs of security and law-enforcement agencies with the needs of others—for ex-

---

Trust in Cyberspace); Computer Science and Telecommunications Board, National Research Council, 2001, *Embedded, Everywhere: A Research Agenda for Networked Systems of Embedded Computers,* National Academy Press, Washington, D.C. (hereafter cited as CSTB, NRC, 2001, *Embedded, Everywhere).*

> **BOX 3.1 Monoculture and System Homogenity**
>
> The existence of "monoculture" on the Internet has both advantages and disadvantages. Its primary advantage is that increased standardization of systems generally allows for greater efficiencies (e.g., easier interoperability). On the other hand, monocultural environments are generally more vulnerable to a single well-designed attack—a fact greatly exacerbated by the extensive interconnections that the Internet provides.
>
> For example, a constant barrage of computer viruses has been designed to attack the weaknesses of the Windows operating system and its associated browser and office productivity programs. However, these viruses have had a negligible direct effect on computers running other operating systems. Furthermore, while these attacks can be propagated through computer servers running other operating systems, they are not propagated *on* these systems.
>
> A monoculture is highly vulnerable to attacks because once a successful attack on the underlying system is developed, it can be multiplied at extremely low cost. Thus, for all practical purposes, a successful attack on one system means that all similar (and similarly configured) systems connected to it can be attacked as well.
>
> As a general rule, inhomogeneity of systems would make broad-based attacks more difficult. This should be considered carefully when designing primary or redundant critical network systems.
>
> (Natural "living systems" provide an interesting analogy to the importance of diversity. Heterogeneity plays an important role in preventing minor changes in the climate, environment, or parasites from destroying the entire system. Different species demonstrate varying levels of vulnerability to the variety of challenges encountered, and this lends resilience to the system as a whole. The diversity of species also lessens the possibility of infectious disease spreading across the entire system.)

ample, the fire and medical personnel on-site—mean that security and law enforcement may interfere with rescue and recovery operations.

3. *Information-fusion systems for the prevention, detection, attribution, and remediation of attacks.* "Information fusion" promises to play a central role in countering future terrorist efforts. Information fusion is an essential tool for the intelligence analysis needed if preemptive disruption of terrorist attacks is to be successful. Knowing that a biological attack is in progress (an issue of detection) or determining the perpetrators of an attack (an issue of attribution) may depend on the fusion of large amounts of information And, in many cases, early warning of an attack increases the effectiveness of any counterresponse to it. In every case, information from many sources will have to be acquired, integrated, and appropriately interpreted to support decision makers (ranging from emergency-response units to intelligence organizations). Given the range of formats, the permanence and growing volume of information from each source, and the difficulty of accurately analyzing information from single let alone multiple sources, information fusion offers researchers a challenge.

It must also be noted that although technology is central to all of the areas listed above, it is not the sole element of concern. Research in these areas must be multidisciplinary, involving technologists, social scientists, and domain experts (e.g., the people from the multiple agencies that need to work together during crisis solutions). All technology deployed for operational purposes is subject to the reality of implementation and operation by humans. Thus, systems issues, including human, social, and organizational behavior, must be part of the research to develop the needed technology and the system design to implement it. Technology cannot be studied in isolation of the ways in which it is deployed, and failure to attend to the human, political, social, and organizational aspects of solutions will doom technology to failure. For this reason, Section 3.6 addresses social and organizational dimensions that must be incorporated into study in these areas.

To assist decision makers in the formulation of a research program, Table 3.1 presents the committee's rough assessment of the criticality of the various research areas identified, the difficulty of the research problems, and the likely time scale on which progress could be made. The criticality of a research area reflects an assessment of the vulnerabilities that might be reduced if significant advances in that area were accomplished and deployed; areas are ranked "High," "Medium," or "Low." How hard it will be to make significant progress is rated "Very Difficult," "Difficult," or "Easy." The time frame for progress is ranked as "1-4 years," "5-9 years," or "10+ years." Of course, the deployment of research results also presents obstacles, which may reduce effectiveness or lengthen the time until a research result can become a reality. It must also be noted that these assessments are both subjective and subject to some debate, as they were intended to provide readers with a quick calibration of these issues rather than a definitive conclusion.

Finally, while R&D is an essential element of the nation's response to counterterrorism, it is not alone sufficient. Indeed, the history of information security itself demonstrates that the availability of knowledge or technology about how to prevent certain problems does not necessarily translate into the application of that knowledge or the use of that technology. It is beyond the scope of this report to address such issues in detail, but policy makers should be cautioned that R&D is only the first step on a long road to widespread deployment and a genuinely stronger and more robust IT infrastructure.

3.1 INFORMATION AND NETWORK SECURITY

A broad overview of some of the history of and major issues in information and network security is contained in the CSTB report *Cybersecurity*

TABLE 3.1 A Taxonomy of Priorities

Category	Research Criticality	Difficulty	Time Scale for R&D for Significant Progress and Deployment
Improved Information and Network Security	High	Difficult	5-9 years
Detection and identification	High	Difficult	5-9 years
Architecture and design for containment	High	Difficult	5-9 years
Large system backup and decontamination	High	Difficult	5-9 years
Less buggy code	High	Very difficult	5-9 years
Automated tools for system configuration	High	Difficult	1-4 years
Auditing functionality	Low	Difficult	10+ years
Trade-offs between usability and security	Medium	Difficult	5-9 years
Security metrics	Medium	Difficult	1-4 years
Field studies of security	High	Easy	1-4 years
C3I for Emergency Response	High	Difficult	1-4 years
Ad hoc interoperability	High	Easy	1-4 years
Emergency deployment of communications capacity	High	Easy	1-4 years
Security of rapidly deployed ad hoc networks	Medium	Difficult	5-9 years
Information management and decision-support tools	Medium	Difficult	5-9 years
Communications with the public during emergency	High	Difficult	1-4 years
Emergency sensor deployment	High	Easy	1-4 years
Precise location identification	Medium	Difficult	5-9 years
Mapping the physical telecommunications infrastructure	High	Easy	1-4 years
Characterizing the functionality of regional networks for emergency responders	High	Difficult	1-4 years
Information Fusion	High	Difficult	1-4 years
Data mining	High	Difficult	1-4 years
Data integration	High	Difficult	1-4 years
Language technologies	High	Difficult	1-4 years
Image and video processing	High	Difficult	5-9 years
Evidence combination	Medium	Difficult	1-4 years
Interaction and visualization	Medium	Difficult	1-4 years
Privacy and Confidentiality	High	Difficult	1-4 years
Planning for the Future	Medium	Difficult	10+ years

Today and Tomorrow: Pay Now or Pay Later.[2] That report builds on a variety of earlier, more detailed CSTB studies related to information and network security.

Despite diligent efforts to create effective perimeter defenses, the penetration of defended computer and telecommunications systems by a determined adversary is highly likely. Software flaws, lax procedures for creating and guarding passwords, compromised insiders, and insecure entry points all lead to the conclusion that watertight perimeters cannot be assumed. Nevertheless, strengthening defensive perimeters is helpful, and this section deals with the methodologies of today and tomorrow that can detect or confine an intruder and, if necessary, aid in recovery from attack by taking corrective action. (Box 3.2 describes some of the fundamental principles of defensive strategy.)

As noted above, the technology discussed here is relevant for efforts in defensive information warfare and for fighting cybercrime. In addition, many advances in information and network security can improve the reliability and availability of computer systems, which are issues of importance to users even under ordinary, nonthreatening conditions. The fact that such advances have dual purposes could help to generate broader interest and support in R&D in this area, as well as to motivate its incorporation into industry products.

Research and development in this area should be construed broadly to include R&D on defensive technology (including both underlying technologies and architectural issues), organizational and sociological dimensions of such security, forensic and recovery tools, and best policies and practices. Research in information and network security can be grouped in four generic areas: authentication, detection, containment, and recovery. A fifth set of topics (e.g., reducing buggy code, dealing with misconfigured systems, auditing functionality) is broadly applicable to more than one of these areas.

3.1.1 Authentication

A terrorist may seek to gain access to a computer system that he or she is not authorized to use. Once access is gained, many opportunities for causing harm are available, including the installation of hostile pro-

---

[2]Computer Science and Telecommunications Board (CSTB), National Research Council (NRC). 2002. *Cybersecurity Today and Tomorrow: Pay Now or Pay Later*. National Academy Press, Washington, D.C. (hereafter cited as CSTB, NRC, 2002, *Cybersecurity Today and Tomorrow*).

> **BOX 3.2 Principles of Defensive Strategy**
>
> Computing and communications systems that contain sensitive information or whose functioning is critical to an enterprise's mission must be protected at higher levels of security than are adequate for other systems. Several policies should be mandatory for such critical systems:
>
> - *The use of encryption for communication between system elements and the use of cryptographic protocols.* These practices help to ensure data confidentiality against eavesdroppers and data integrity between major processing elements (e.g., host to host, site to site, element to element); prevent intrusion into the network between nodes (e.g., making "man-in-the-middle" attacks much more difficult); and provide strong authentication (e.g., through the use of public-key-based authentication systems that use encryption and random challenge to strengthen the authentication process or to bind other elements of the authentication, such as biometrics, to the identity of a user).
> - *Minimal exposure to the Internet, which is inherently insecure.* Firewalls provide a minimal level of protection, but they are often bypassed for convenience. (Balancing ease of use and security is an important research area discussed in the main text of this report.) Truly vital systems may require an "air gap" that separates them from public networks. Likewise, communication links that must remain secure and available should use a private network. (From a security perspective, an alternative to a private network may be the use of a connection on a public network that is appropriately secured through encryption. However, depending on the precise characteristics of the private network in question, a private network may—or may not—provide higher availability.)
> - *Strong authentication technology for authenticating users.* Security tokens based on encryption (such as smart cards) are available for this purpose, and all entrances from a public data network (such as a network access provider or insecure dial-in) should use them. Furthermore, for highly critical systems, physical security must also be assured.
> - *Robust configuration control.* Such control is needed to ensure that only approved software can run on a system and that all of the security-relevant knobs and switches are correctly set.
>
> Such measures are likely to affect ease of use and convenience, as well as cost. These are prices that must be paid, however, because hardening critical systems will greatly reduce vulnerability to a cyberattack.

grams and the destruction or compromise of important data. In other instances, a terrorist might orchestrate the actions of multiple computers to undertake harmful acts, for example through denial-of-service attacks.[3]

---

[3]Computer Science and Telecommunications Board (CSTB), National Research Council (NRC), 1999, *Realizing the Potential of C4I: Fundamental Challenges,* National Academy Press, Washington D.C., pp. 144-152 (hereafter cited as CSTB, NRC, 1999, *Realizing the Potential of C4I*); CSTB, NRC, 1999, *Trust in Cyberspace.*

To prevent a terrorist from gaining unauthorized access to a computer, it is necessary to prevent that person from successfully posing as an authorized user.[4] In other words, an authorized user must pass successfully through an authentication process that confirms his or her asserted identity as an authorized user. The same is true of computer-to-computer interactions—for some transactions, it is necessary for Computer A to determine if Computer B is one of the computers authorized to interact with it. Here too, a computer-to-computer authentication process is necessary so that only individual authorized devices connecting to a network can receive services.

Today, the prevailing method of user authentication is the password. Passwords are easily compromised by the use of weak password-choosing techniques, the recording of passwords in open places, the communication channel used for the password entry or administration, and by password-cracking techniques. Requiring several log-in procedures (and usually with different passwords for each) for mapping to permission or authorization techniques makes system administration even more complex for the user and the system administrator. Other devices, such as hardware tokens, are usually more secure than are passwords, but a user must have them available when needed.

The ideal authentication system would be a simple, easy-to-use system that verified identity, could be managed in a distributed manner, had the trustworthiness of cryptographically based systems without today's complexity, could be scaled to hundreds of thousands (or even millions), and had a cost of ownership that was competitive with passwords. In practice, these desirable attributes often entail trade-offs with one another; one way to focus a research effort in authentication would be to address the reduction of these trade-offs.[5]

3.1.2 Detection

Even with apparently secure authentication processes and technologies, it might still be possible for an intruder to gain unauthorized access to a system, though with more effort if the system were more secure. This possibility suggests a need for detecting and identifying intruders. How-

---

[4]The case of a terrorist as an insider who has already been granted access is not within the scope of this particular problem. Insider attacks are addressed in other parts of this report.

[5]An ongoing CSTB project examines in detail the technologies underlying authentication. See the Web site <http://www.cstb.org> for more information on this subject.

ever, intruders are often indistinguishable from valid users and frequently take great care to hide their entry and make their behavior look innocuous.

Intrusion-detection systems are designed to monitor users and traffic to detect either anomalous users or unusual traffic patterns that might indicate an active attack. Of course, such monitoring requires good characterizations of what "normal" behavior is and knowledge of what various kinds of behavior mean in the context of specific applications. Today, the major deficiency in this approach is the occurrence of too many false positives. That is, the behavior of legitimate users is sufficiently diverse that some types of legitimate behavior are mischaracterized as anomalous (and hence hostile). Thus, research is needed on reducing the rate of false positives in intrusion-detection systems.

Another approach to intrusion is based on deceiving the cyber-attacker. For example, traps (sometimes referred to as honeypots)—such as apparently interesting files—can be crafted to attract the attention of an intruder so that he or she might spend extra time examining it. That extra time can then be used to provide warning of hostile intent, and it might help in forensic investigation while the hostile party is connected to the system. Alternatively, tools might be created that disguise the actual details of a network when it is probed. Tools of this nature, as well as the development of forensic tools for use in attacker-deceiving environments, may be fruitful areas of research.

A related challenge is the development of intruder-detection methods that scale to function efficiently in large systems. Current approaches to intrusion detection generate enormous amounts of data; higher priority must be given to systems that can analyze rather than merely collect such data, while still retaining collections of essential forensic data. Moreover, the collection and analysis of such large amounts of data may degrade performance to unacceptable levels.

Intrusion-detection systems are also one element of technology that can be used to cope with the threat of insider attack. Because the trusted insider has legitimate access to system resources and information, his or her activities are subject to far less suspicion, which usually allows the insider to act with much greater freedom than would be permitted an outsider who had penetrated the system. Thus, new technologies specifically focused on the possibility of insider attack may be a particularly fruitful avenue of research (Box 3.3). In addition, research focused on understanding common patterns of insider attack (e.g., through the use of application-level audits to examine usage patterns) could be integrated with other kinds of audits to provide a more robust picture of system usage.

> **BOX 3.3 Illustrative Technologies for Dealing with the Insider Attack**
>
> Authentication, access control, and audit trails are three well-understood technologies that can be used in combating the insider threat. Using these mechanisms to enforce strict accountability can improve protection against the insider threat, but in practice they are often not as successful as they might be. For example, many current tools for access control and audits are difficult to use, or they generate large volumes of data that are, for practical purposes, unreviewable.
>
> Other technology research areas that may be relevant to dealing with insider attack include:
>
> - *Attack-specification languages.* Programming languages designed for ease of modeling attacks and/or expressing attack behaviors and modalities.
> - *Modeling and simulation of insider attacks.* Better understanding of such attacks to help those seeking to validate technologies to counter the insider threat. Today, simulations of such attacks are difficult to perform and are personnel-intensive.
> - *Authentication of roles, rights, privileges.* Approaches to using finer-grained authentication strategies based not on authenticating an individual as an individual but as the holder of certain rights and privileges or embodying a certain role.
> - *Semantics of authorized access.* Development of the semantics of operations and authorization that enable more fine-grained authorization decisions or the flagging of potentially suspect audit trails.
> - *Automated, dynamic revocation of privileges.* Development of effective strategies for the automatic revocation of privileges based on policy-specified factors such as the timing of certain types of access or other user behavior.
> - *Fingerprinting of documents.* Strategies for embedding identifying information into a document so that its subsequent disposition can be more effectively traced. Such information would be analogous to a copyright notice in a document that can help to determine its origin, except that it would not be easily removable from the document itself.
> - *Continuous authentication.* Technologies for authenticating a user after the initial authentication challenge (to deal with the fact that people walk away from their computers without logging out).

3.1.3 Containment

Today's systems and networks often fail catastrophically. That is, a successful attack on one part of a system can result in an entire system or network's being compromised. (For example, the failure of a perimeter defense, such as a firewall, surrounding otherwise unprotected systems can result in an intruder's gaining full and complete access to all of those systems.) A system that degrades gracefully is more desirable—in this case, a successful attack on one part of a system results only in that part's being compromised, and the remainder of the system continues to function almost normally.[6]

The principle of graceful degradation under attack is well accepted, but system and network design for graceful degradation is not well understood. Nor are tools available to help design systems and networks in such a manner. Even more difficult is the challenge of modifying existing legacy systems to fail gracefully.

In addition, the building blocks of today's systems are generally commercial off-the-shelf components.[7] Despite the security limitations of such components, the economics of system development and the speed with which the IT environment changes inevitably require them to be built this way. However, it is not known today how to integrate components safely, how to contain faults in them, and how to disaggregate them when necessary. While this lack of understanding applies to systems ranging from accounting and payroll systems to telephone switching systems, SCADA systems are a particularly important case.

Architectural containment as a system-design principle calls for the ability to maintain critical functionality (such as engine control on a ship) despite failures in other parts of a system.[8] A sophisticated control system used during "normal operations" must be able to provide basic functionality even when parts of it have been damaged.[9] Such an approach could be one of the most effective long-term methods for hardening IT targets that oversee critical operations.

For the most part, current approaches to system design involve either the independence of system components (which in modern large-scale systems leads to inefficiencies of operation) or the integration of system components (with the inherent vulnerabilities that this approach entails). Containment essentially navigates between the two extremes; its essential element is the ability to "lock down" a system under attack—perhaps to suspend normal operation temporarily while preserving some basic functionality as the system finds and disables potential intruders and to resume normal system operation afterward—with less disruption than might be caused by shutting down and rebooting.

Research is thus necessary in several areas: in understanding how to

---

[6]CSTB, NRC, 1999, *Realizing the Potential of C4I*, pp. 144-152.

[7]CSTB, NRC, 1999, *Trust in Cyberspace*.

[8]It should be noted that an essential aspect of designing for containment is the ability to define and prioritize which functions count as essential. For systems used by multiple constituencies, this ability cannot be taken for granted.

[9]As an example, a shipboard networking failure on the USS *Yorktown* left the ship without the ability to run its engines. (Gregory Slabodkin. 1998. "Software Glitches Leave Navy Smart Ship Dead in the Water." *Government Computing News*, July 13.)

fuse a simple, highly secure, basic control system used primarily for operations under attack with a sophisticated, highly effective, and optimized control system used for normal operations; in the "decontamination" of a system while it is being used (see Section 3.1.4); and in the resumption of operations without the need for going offline. One "grand challenge" might be the development of a system that could be made more secure at the touch of a button; the cost would be the loss of some nonessential functionality while the system simultaneously decontaminated itself or shut out attackers. A second challenge involves existing systems that need to be examined and restructured. Methods and tools for analyzing systems to identify essential functionality and to restructure those systems to tolerate the loss of nonessential functions could be productively applied to existing designs well before a body of design principles and theories become clear.

A serious problem for which few general solutions are known is the denial-of-service (DOS) attack. For example, consider a DOS attack that is launched against the major Internet news services to coincide with a physical bomb attack on some other target such as a crowded sports stadium. It would be nearly impossible to distinguish legitimate users, who would simply be looking for information, from attackers inundating the news service Web sites to try to prevent access to that information, possibly increasing panic and spreading misinformation. A *distributed* denial-of-service (DDOS) attack uses many computers to launch attacks against a given target, rendering ineffective the obvious approach of simply cutting off an attacking computer.

One approach to countering a DOS attack calls for authentication so that intruders and bogus traffic can more easily be distinguished. Developing such methods that are both fast and scalable (i.e., effective and fast even when they involve the authentication of large numbers of parties) remains the major challenge in this area, however. (A technique that may be worth further development, at least in the context of authenticating traffic to and from heavily used Web sites, is easy-to-use subscription models.[10]) In any event, research on countering DOS attacks is important.

Another approach within this general area of containment is the de-

---

[10]A subscription model calls for a user to register for service in some authenticated way, so that a site can distinguish that user from a random bad user. Because denial-of-service attacks depend on a flood of bogus requests for service, the availability of a database of registered users makes it easy to discard service requests from requests that are not registered—and those are likely to account for the vast majority of bogus requests. (Of course, there is nothing in this scheme to prevent a registered user from conducting a hostile action, but the number of hostile registered users is likely to be small, allowing counteractions to be taken on an individual basis.)

velopment of broad architectural principles of robust infrastructure for different kinds of applications (e.g., SCADA systems). For example, principles are needed to guide decisions about the relative robustness of distributed power generation versus centralized power generation in the face of attack.

3.1.4 Recovery

Once an intruder has been detected, confined, and neutralized, the goal becomes recovery—the restoration of a system's full functional operating capability as soon as possible. As with containment, recovery is highly relevant for reliability, although the presence of a determined adversary makes the problem considerably harder. Recovery includes preparations not only to help ensure that a system is recoverable but also to be able to actively reconstitute a good system state.

Backup is an essential prerequisite for reconstitution. Although the basic concepts of system backup are well understood, there are major challenges to performing and maintaining backups in real time so that as little system state as possible is lost. However, normal backup methods have been developed under the assumption of benign and uncorrelated failure, rather than that of a determined attacker who is trying to destroy information. Further, backups of large systems take a long time, and if they are in use during the backup, the system state can change appreciably during that time. Thus, research is needed on ways to preserve information about system state during backup.

Reconstituting a computer system after an intrusion relies on the use of backup. Unlike a restore operation used to re-create a clean system after a failure, reconstitution requires an additional step: decontamination, which is the process of distinguishing between clean system state (unaffected by an intruder) and the portions of infected system state, and eliminating the causes of those differences. Because system users would prefer that as little good data as possible be discarded, this problem is quite difficult. Decontamination must also remove all active infections, as well as any dormant ones. Once decontamination is performed, attention can be turned to forensics in an attempt to identify the attacker[11] and acquire evidence suitable for prosecution or retaliation. In the end, this ability is critical to long-term deterrence.

Given that penetration of computer and telecommunications networks is likely to continue despite our best efforts to build better perimeter security, more resilient and robust systems are necessary, with backup and recovery as essential elements.

---

[11]CSTB, NRC, 1999, *Realizing the Potential of C4I*, pp. 144-152.

New approaches to decontamination are also needed, especially when a system cannot be shut down for decontamination purposes. At present, much of the activity associated with a properly running system interferes with decontamination efforts (particularly with respect to identifying a source of contamination and eliminating it).

3.1.5 Cross-cutting Issues in Information and Network Security Research

A number of issues cut across the basic taxonomy of detection, containment, and recovery described above.

Reducing Buggy Code

Progress in making systems more reliable will almost certainly make them more resistant to deliberate attack as well. But buggy code underlies many problems related both to reliability and to security, and no attempt to secure systems and networks can succeed if it does not take this basic fact into account.[12]

Buggy software is largely a result of the fact that, despite many years of serious and productive research in software engineering, the creation of software is still more craft than science-based engineering. Furthermore, the progress that has been made is only minimally relevant to the legacy software systems that remain in all infrastructures.

The fact that essentially all software systems contain bugs is not new. Bugs can result from a variety of causes, ranging from low-level errors (e.g., a mathematical expression uses a plus sign when it should use a minus sign) to fundamental design flaws (e.g., the system functions as it was designed to function, but these functions are inappropriate in the circumstances of operation).

Dealing with buggy code is arguably the oldest unsolved problem in computer science, and there is no particular reason to think that it can be solved once and for all by any sort of crash project. Nevertheless, two areas of research seem to be particularly important in a security context:

- *Security-oriented tools for system development.* More tools that support security-oriented development would be useful.[13] For example,

---

[12]CSTB, NRC, 1991, *Computers at Risk*; CSTB, NRC, 1999, *Trust in Cyberspace*.

[13]Catherine Meadows, 1996, "The NRL Protocol Analyzer: An Overview," *Journal of Logic Programming*, Vol. 26 (2); Dawn Song et al., 2001, "Athena, a Novel Approach to Efficient Automatic Security Protocol Analysis," *Journal of Computer Security*, Vol. 9 (1, 2).

model-checking tools, which have been used successfully for hardware verification, can be used for analyzing designs of security protocols including authentication protocols[14] and electronic commerce protocols;[15] static analyzers that have been used successfully for compiler optimization may be usable to analyze code for information flow properties;[16] dynamic analyzers for online monitoring (detection) and reconfiguring (response and recovery) may be used for intrusion detection and forensics;[17] and theorem-proving tools[18] that have been used for proof-carrying code[19] can be extended to handle more expressive security-specific logics.

- *Trustworthy system upgrades and bug fixes.* It often happens that a system bug is identified and a fix to repair it is developed. Obviously, repairing the bug may reduce system vulnerability, and so system administrators and users—in principle—have some incentive to install the patch. However, with current technology, the installation of a fix or a system upgrade carries many risks, such as a nontrivial chance of causing other problems, a disruption of existing functionality, or possibly the creation of other security holes, even when the fix is putatively confined to a module that can be reinstalled.[20] The essential reason for this problem is

---

[14]Gavin Lowe, 1996, "Breaking and Fixing the Needham-Schroeder Public-Key Protocol Using FDR," in *Tools and Algorithms for the Construction and Analysis of Systems*, Tiziana Margaria and Bernard Steffen (eds.), Vol. 1055 of *Lecture Notes in Computer Science*, Springer Verlag; Will Marrero et al., 1997, *Model Checking for Security Protocols*, Technical Report 97-139, Department of Computer Science, Carnegie Mellon University, Pittsburgh, Pa., May; J.C. Mitchell et al., 1997, "Automated Analysis of Cryptographic Protocols Using Murphi," IEEE Symposium on Security and Privacy, Oakland, available online at <http://theory.stanford.edu/people/jcm/papers/murphi-protocols.ps>.

[15]Nevin Heintze et al., 1996, "Model Checking Electronic Commerce Protocols" (extended abstract), USENIX Workshop on Electronic Commerce; Darrell Kindred and Jeannette M. Wing, 1996, "Fast, Automatic Checking of Security Protocols," in *Proceedings of the USENIX 1996 Workshop on Electronic Commerce*, November.

[16]Andrew C. Myers and Barbara Liskov. 1997. "A Decentralized Model for Information Flow Control," in *Proceedings of the ACM Symposium on Operating System Principles (SOSP '97)*. Saint Malo, France, October.

[17]Giovanni Vigna and Richard A. Kemmerer. 1999. "NetSTAT. A Network-based Intrusion Detection System." *Journal of Computer Security*, Vol. 7 (1).

[18]Lawrence C. Paulson. 1999. "Proving Security Protocols Correct," pp. 370-381 in *IEEE Symposium on Logic in Computer Science*. Trento, Italy. Available online at <http://www.cl.cam.ac.uk/users/lcp/papers/Auth/lics.pdf>.

[19]George C. Necula and Peter Lee. 1997. "Proof-Carrying Code," pp. 106-119 in *Proceedings of the 24th Annual ACM SIGPLAN-SIGACT Symposium on Principles of Programming Languages (POPL)*. ACM Press, New York, January.

[20]Frederick P. Brooks. 1975. *The Mythical Man-Month*. Addison-Wesley, Boston, Mass.

that although fixes are tested, the number of operational configurations is much larger than the number of test configurations that are possible. Research is thus needed to find ways of testing bug fixes reliably and to develop programming interfaces to modularize programs that cannot be bypassed. An additional dimension of trustworthy upgrades and bug fixes is their installation. If such upgrades and fixes can be developed, their automatic online installation becomes a reasonable desire. Today, some operating systems and programs allow automatic download of such modifications, but it is often the case that they cannot be installed while the system is running. Moreover, in light of the security concerns mentioned in Section 2.2.6, automatic downloads of bug fixes must be made very highly secure. Research will be needed to solve this problem.[21]

Misconfigured Systems (Configuration Management)

Because existing permission and policy mechanisms are hard to understand, use, and verify, many system vulnerabilities result from their improper administration.[22] There is also a trade-off between fine-grained access control and usability, and both needs are growing. For example, an entire group of people may be given access privileges when only one person in that group should have them. Or, a local system administrator may install a modem on the system he or she administers with the intent of obtaining access from home, but this also provides intruders with an unauthorized access point.

The ability to formulate security policies at the appropriate level, to ensure that policies are consistent across all levels and to state these policies crisply and clearly (e.g., what language is used to express the policy, what mechanisms are used to enforce it, and what is to be done if it is violated) would be helpful. In addition, it will be necessary to develop methods of generating formal descriptions of actual, existing security policies in place and to compare them with desired security policies. Thus, better tools for formulating and specifying security policies and for checking system configurations quickly against prespecified configurations should be developed. Better tools for system and network operators to detect added and unauthorized functionality (e.g., the addition of a Trojan horse) are also necessary.

---

[21] Advances in this area would also have substantial benefit in reducing the workload of system administrators, who today must spend considerable time keeping up with the sheer volume of security patches and assessing the costs and benefits of installing them.

[22] CSTB, NRC, 1991, *Computers at Risk*; CSTB, NRC, 1999, *Trust in Cyberspace*.

Auditing Functionality

Validation sets are used to ensure that a piece of hardware (e.g., a chip) has the functionality that its design calls for. However, these sets typically test for planned functionality—that is, can the hardware properly perform some specified function? They do not test for unauthorized functionality that might have been improperly inserted, perhaps by someone seeking to corrupt a production or distribution chain. Similar problems are found in the development of software, especially software systems with many components that are developed by multiple parties. Research is needed for developing tools to ensure that all of the called-for functionality is present *and* that no additional functionality is present. Research is needed to develop tools that can detect added unauthorized functionality.

Managing Trade-offs Between Functionality and Security

As a general rule, more secure systems are harder to use and have fewer features.[23] Conversely, features—such as executable content and remote administration—can introduce unintended vulnerabilities even as they bring operational benefits. (For example, newer word processors allow the embedding of macros into word processing files, a fact that results in a new class of vulnerabilities for users of those programs as well as added convenience.)

One good example of the trade-off between security and usability is the difficulty of establishing an encrypted communications channel using a popular encryption program. One study found that a majority of test-population individuals experienced in the use of e-mail were unable to sign and encrypt a message using PGP 5.0. In a security context, these usability issues go beyond the user-interface design techniques appropriate to most other types of software.[24]

A second example is the buffer overflow problem.[25] Over half of the

---

[23]CSTB, NRC, 1991, *Computers at Risk*, pp. 159-160.

[24]For example, from the user's perspective, security is secondary to the primary goal of using the software for some practical purpose. Traditional user-interface design techniques presume that users are motivated to use software for their primary purposes. More discussion of this and other relevant issues can be found in Alma Whitten and J.D. Tygar. 1999. "Why Johnny Can't Encrypt: A Usability Evaluation of PGP 5.0," *Proceedings of the 9th USENIX Security Symposium*, August.

[25]In a buffer overflow, memory is overwritten by an application and control is transferred to rogue code. Type-safe languages allow memory accesses only to specifically authorized locations. For example, programs written in type-safe languages cannot read or write to memory locations that are associated with other programs.

system and network security vulnerabilities documented by the Computer Emergency Response Team (CERT) of the Software Engineering Institute at Carnegie Mellon University throughout CERT's existence involve buffer overflows. These can exploited by an adversary to gain control over a target system. Some ways of preventing buffer overflows are known, but they involve degradations in performance. For example, it is possible for processors to support a programmer-usable capability to designate certain areas of memory as "code" (that cannot be overwritten) versus "data" (that can be overwritten). If this designation is made, buffer overflows cannot result in the transfer of control to rogue code. On the other hand, the extra hardware needed to implement this capability has the effect of slowing down memory access by nontrivial amounts. Java and similar type-safe languages are also more resistant to buffer overflows than are other languages, but there is overhead in undertaking the check of parameter ranges needed to prevent overflows.

More research is required for performing essential trade-offs between a rich feature set or performance and resistance to attack. Transparent security would be more acceptable to users and hence would be employed more frequently. New authentication mechanisms (or implementations) that combine higher security with lower inconvenience are also needed.

Security Metrics

Many quantitative aspects of security are not well understood. For example, even if a given security measure is installed—and installed properly (something that cannot be assumed in general)—there is no way of knowing the degree to which system security has increased. Threat models are often characterized by actuarial data and probability distributions in which the adverse effects of vulnerabilities are prioritized on the basis of how likely they are to occur; but such models are of little use in countering deliberate terrorist attacks that seek to exploit nominally low-probability vulnerabilities. Notions such as calculating the return on a security investment—common in other areas in which security is an issue—are not well understood either, thus making quantitative risk management a very difficult enterprise indeed.[26] Research is needed for developing meaningful security metrics.

---

[26]Information on the economic impact of computer security is available online at <http://www.nist.gov/director/prog-ofc/report02-1.pdf>.

Intelligence Gathering

Given the rate at which information technology changes, it is likely that new vulnerabilities and new types of attack will emerge rapidly. Because insight into the nature of possible attacks is likely to result in additional options for defense, it is highly desirable to keep abreast of new vulnerabilities and to understand the potential consequences if such vulnerabilities were to be exploited.

Field Studies of Security

Traditional criteria for secure systems, as specified in the "Orange Book,"[27] have not been successes. They do not capture current needs or models of computation.[28] Worse yet, they have largely failed in the marketplace; very few customers actually bought Orange Book-rated systems, even when they were available.[29] Understanding why previous attempts to build secure systems and networks have failed in the marketplace, or in defending against outside attack, would help to guide future research efforts. (Further, human and organizational factors are key elements of such analysis, as previously described, and the understanding from such study is likely to be based at least as much on economics, sociology, and anthropology as on technology.[30])

3.2 SYSTEMS FOR EMERGENCY RESPONSE

Technologies for command, control, communications, and intelligence (C3I) have major importance in the response phase of a disaster. In general, the IT infrastructure for emergency responders must be robust in the

---

[27]The "Orange Book" is the nickname for *Trusted Computer System Evaluation Criteria*—criteria that were intended to guide commercial system production generally and thereby improve the security of systems in use. See U.S. Department of Defense. 1985. *Trusted Computer System Evaluation Criteria*. Department of Defense 5200.28-STD, "Orange Book." National Computer Security Center, Fort Meade, Md., December.

[28]CSTB, NRC, 1999, *Realizing the Potential of C4I*, pp. 144-152; CSTB, NRC, 1991, *Computers at Risk*; CSTB, NRC, 1999, *Trust in Cyberspace*.

[29]For example, commercial needs for computer security focused largely on data integrity, while military needs for security focused on confidentiality, as noted in David Clark and David Wilson, 1987, "A Comparison of Commercial and Military Computer Security Policies," in *Proceedings of the 1987 IEEE Symposium on Security and Privacy*, IEEE, Oakland, Calif. Another key failing of the Orange Book approach to security included its omission of networking concerns. For more discussion, see CSTB, NRC, 1999, *Trust in Cyberspace*.

[30]See, for example, Donald Mackenzie. 2001. *Mechanizing Proof: Computing, Risk, and Trust*. MIT Press, Cambridge, Mass.

face of damage and even potential terrorist attack.[31] Although the incident management process has been well studied,[32] the IT requirements for such management do not appear to have been thoroughly conceived, even though in a disaster it is essential that IT systems provide for the capability to deliver information, interagency communication and coordination, and communication with those affected both within and beyond the immediate disaster area. In a disaster, equipment must be deployed immediately to provide for appropriate communication to those responding to the situation, and it must be deployed to the multiple agencies in the private and public sectors that are affected, and to and between those directly affected by the incident.[33]

The committee believes that research in a number of areas, described below, can advance the state of the art for C3I systems for emergency response (and provide collateral benefits for the responses to more common natural disasters as well).

3.2.1 Intra- and Interoperability

The C3I systems of emergency-response agencies must support both intra-agency and interagency communications. But many disasters (whether natural or attack-related) reveal technological shortcomings in a given agency's C3I systems as vital communications are lost or never heard. And, although the public rhetoric of every emergency-response agency acknowledges the need for cooperation with other agencies, almost every actual disaster reveals shortcomings in the extent and nature of interagency cooperation.

Perhaps the most basic requirement for an agency's C3I systems is that they reliably support communications among the personnel of that agency. The C3I systems of most emergency responders today are based on analog radio technology. Although digital and analog communica-

---

[31]Computer Science and Telecommunications Board (CSTB), National Research Council (NRC). 1999. *Information Technology Research for Crisis Management*. National Academy Press, Washington, D.C., p. 39 (hereafter cited as CSTB, NRC, 1999, *Information Technology Research for Crisis Management*).

[32]Hank Christen, Paul Maniscalco, Alan Vickery, and Frances Winslow. 2001. "An Overview of Incident Management." *Perspectives on Preparedness*, No. 4, September. Available online at <http://ksgnotes1.harvard.edu/BCSIA/Library.nsf/pubs/POP4>. Accessed November 14, 2002.

[33]Computer Science and Telecommunications Board (CSTB), National Research Council (NRC). 1996. *Computing and Communications in the Extreme: Research for Crisis Management and Other Applications*. National Academy Press, Washington, D.C., p. 14 (hereafter cited as CSTB, NRC, 1996, *Computing and Communications in the Extreme*).

tions systems have advantages and disadvantages (Box 3.4), the disadvantages of analog communications for emergency responders are considerable in the context of large-scale disasters or incidents.

For most routine work, the mostly-analog systems deployed today function adequately to keep personnel in contact with one another. And, because routine work is mostly what response agencies do, the motivation for acquiring new systems is low. Even when new procurements are considered, the needs highest on an agency's list of priorities are most likely to be those that focus on enhancing its ability to do its everyday work more effectively.

Nevertheless, disasters increase the demand for communications among emergency responders dramatically, and the capacity limitations of analog systems become more apparent. In addition, large-scale disasters may affect telecommunications facilities. For example, antennas supporting the communications systems of the Fire Department of New York, the New York Police Department, and the Emergency Medical System (EMS) were based on the roof of Building 1 of the World Trade Center, which collapsed soon after the strike.[34]

Inadequacies in the deployed base of C3I systems for emergency responders cannot be fixed merely by adding more channels. Though emergency planning often results in the allocation of additional channels, these channels ease the problem only temporarily. In addition, extra channels require more bandwidth, and hence a broader radio spectrum dedicated to public service communication, even though radio spectrum is a limited resource. More efficient use of spectrum can be obtained by using digital communications systems, which offer additional advantages as well (Box 3.4), such as greater noise immunity and higher security against eavesdropping (especially with encryption). However, one major impediment to the acquisition of such systems is inadequate funding. Funding needs are further exacerbated by the difficulty of acquiring new digital systems that preserve backwards compatibility with existing legacy analog systems. (In New York City, the Fire Department relied on radios that were at least 8 years old and in some cases 15 years old, and a senior Fire Department official reported that "there [are] problems with the radios at virtually every high-rise fire."[35])

---

[34]Ronald Simon and Sheldon Teperman. 2001. "Lessons for Disaster Management." *Critical Care*, Vol. 5:318-320. Available online at <http://www.disasterrelief.org/Disasters/011115wtclessons/>.

[35]Jim Dwyer, Kevin Flynn, and Ford Fessenden. 2002. "9/11 Exposed Deadly Flaws in Rescue Plan." *New York Times*, July 7. Available online at <http://www.nytimes.com/

> **BOX 3.4 A Comparison of Digital and Analog Wireless Communications Technologies**
>
> *Advantages of Digital Communications*[1]
>
> • Digital communications provide higher fidelity and lower susceptibility to interference and static compared with analog communications.
> • Digital communications are more inherently secure than analog communications are, and more easily encrypted as well.
> • Channels (frequencies) can be shared between users, thus increasing the number of supportable users per channel.
> • Because channels can be shared, multiple users can speak simultaneously on a single frequency, thus increasing the amount of information that can be exchanged.
>
> *Advantages of Analog Communications*
>
> • Analog communications devices are easier to make interoperable because it is only necessary to match the frequencies of communicating wireless sets, whereas digital systems require the additional step of matching their communications protocols.
> • Problems in analog communications devices are often easier to diagnose, and communications problems in analog systems are often easier to work around than are those in digital systems.
> • The number of technicians and users trained in the use of analog systems is larger.
> • Analog communications offer better graceful degradation than that offered by digital systems in the presence of noise as signal-to-noise ratios drop, and they offer better warning to users (through the tone of the communications carried) when they are near the margins of usability.
> • Low-frequency transmissions, which are better suited for analog communications than for digital communications, are less subject to line-of-sight restrictions and thus have a higher likelihood of penetrating most walls and bypassing debris.
>
> ---
>
> [1]Viktor Mayer-Schönberger. 2002. *Emergency Communications: The Quest for Interoperability in the United States and Europe.* BCSIA Discussion Paper 2002-7. John F. Kennedy School of Government, Harvard University, Cambridge, Mass., March.

2002/07/07/nyregion/07EMER.html?pagewanted=1>. Also according to this article, the New York Fire Department had replaced at least some of its analog radios in early 2001 with digital technology better able to transmit into buildings; but after a few months, these new radios were pulled from service because several firefighters said they had been unable to communicate in emergencies. It is not clear whether or not a full "head-to-head" systematic comparison between the analog and digital systems was ever undertaken.

Interoperability is a broad and complex subject rather than a binary attribute of systems, and it is important to distinguish between interoperability at the operational and technical levels.[36] Operational interoperability refers to the ability of different operating agencies (e.g., police, fire, rescue, utilities) to provide information services to and accept information services from other agencies and to use these services in support of their operational goals. Thus, the dimensions of operational interoperability go beyond IT systems to include people and procedures, interacting on an end-to-end basis. By contrast, technical interoperability refers to the ability of IT systems to exchange information or services directly and satisfactorily between them and/or their users. Technical interoperability involves the ability to exchange relevant bitstreams of information and to interpret the exchanged bits according to consistent definitions—merely providing information in digital form does not necessarily mean that it can be readily shared between IT systems.

Furthermore, numerous computational and database facilities must be established to provide complete and real-time information[37] to diverse constituencies whose information and communication requirements, security needs, and authorizations all differ. These facilities must be established quickly in disaster situations, as minutes and even seconds matter in the urgent, early stages of an incident.[38] Furthermore, tight security is essential, especially if the incident is the result of a terrorist attack, because an active adversary might try to subvert the communications or destroy data integrity.[39] In addition, an atmosphere of crisis and emergency provides opportunities for hostile elements to overcome security measures that are normally operative under nonemergency circumstances. Thus, another research area is how to build systems that permit security exceptions to be declared without introducing new vulnerabilities on a large scale.[40]

---

[36] An extended discussion of interoperability, though in a military context, is provided in CSTB, NRC, 1999, *Realizing the Potential of C4I*.

[37] CSTB, NRC, 1999, *Information Technology Research for Crisis Management*, p. 29.

[38] CSTB, NRC, 1999, *Information Technology Research for Crisis Management*, p. 83; CSTB, NRC, 1996, *Computing and Communications in the Extreme*, p. 12.

[39] CSTB, NRC, 1996, *Computing and Communications in the Extreme*, p. 24.

[40] For example, in a crisis, emergency responders may need to identify all of the individuals working at a specific company location, and tax records might well be one source of data to construct such a list. However, while names and home addresses may be relevant, income is almost certainly not relevant. Technology that enables controlled release under exceptional circumstances and a means to discern when an "exception state" exists would be helpful for such scenarios, although implementations that simply weaken access control policies in an emergency are highly vulnerable to improper compromise through social

Efforts to coordinate communications are complicated by the fact that emergency response to a large-scale incident has many dimensions, including direct "on-the-ground" action and response, management of the incident response team, operations, logistics, planning, and even administration and finance. Moreover, response teams are likely to include personnel from local, county, state, and federal levels.[41]

Poor interoperability among responding agencies is a well-known problem (see, for example, Box 3.5)—and one that is as much social and organizational as it is technical.[42] The fundamental technical issue is that different agencies have different systems, different frequencies and waveforms, different protocols, different databases, and different equipment.[43] (Box 3.6 addresses some of the impediments to interoperability.) Moreover, existing interoperability solutions are ad hoc and do not scale well.[44] The nature of agencies' missions and the political climates in which they traditionally operate make it even harder for them to change their communication methods. Thus, it is unlikely that agencies will ever be strongly motivated to deploy interoperable IT systems. Efforts to solve problems of interagency cooperation by fiat are likely to fail to achieve their goal unless they address interagency rivalries and political infighting about control and autonomy. Efforts to achieve interoperability somehow must work within this reality of organizational resistance (as discussed further in Section 3.6.3).[45]

For practical purposes, these comments suggest that when large, multiagency responses are called for, the individual "come-as-you-are" communications systems will eventually have to be transitioned to an interoperable structure that supports all agencies involved—and transitioned with minimal disruption of function during that transfer.[46] This complex problem has technological and social dimensions—the tech-

---

engineering (see Section 3.6.1). By contrast, there are fewer risks of abuse (but also less flexibility) in implementations that are based on a formal declaration of emergency or a particular threat condition rather than on a conversation persuading an operator that such a situation exists. For more discussion, see Computer Science and Telecommunications Board (CSTB), National Research Council (NRC), 2002, *Information Technology Research, Innovation, and E-Government*, National Academy Press, Washington, D.C. (hereafter cited as CSTB, NRC, 2002, *Information Technology Research, Innovation, and E-Government*).
[41]CSTB, NRC, 1999, *Information Technology Research for Crisis Management*, p. 7.
[42]See for example, Viktor Mayer-Schönberger, 2002, *Emergency Communications: The Quest for Interoperability in the United States and Europe*, BCSIA Discussion Paper 2002-7, John F. Kennedy School of Government, Harvard University, Cambridge, Mass., March.
[43]CSTB, NRC, 1999, *Information Technology Research for Crisis Management*, p. 26.
[44]CSTB, NRC, 1996, *Computing and Communications in the Extreme*, p. 119.
[45]CSTB, NRC, 1999, *Information Technology Research for Crisis Management*, p. 27.
[46]CSTB, NRC, 1999, *Information Technology Research for Crisis Management*, p. 26.

> **BOX 3.5 Examples of Interoperability Difficulties Among Emergency Responders**
>
> *Columbine High School*
>
> In April 1999, two 16-year-old students entered Columbine High School in Littleton, Colorado, and started a shooting spree that left 15 people dead and dozens of others wounded.[1] Within minutes of the first shootings, local police, paramedics, and firefighters arrived at the scene. Over the next several hours, they were joined by almost 1,000 law-enforcement personnel and emergency responders. These responders included police forces from 6 sheriff's offices and 20 area police departments, 46 ambulances and 2 helicopters from 12 fire and emergency medical service agencies, as well as personnel from a number of state and federal agencies. However, there was no communication system that would permit the different agencies to communicate with one another, and thus coordination became a serious problem. Agencies used their own radio systems, which were incompatible with other systems. With more and more agencies arriving on the scene, even the few pragmatic ways of communication that had been established, such as sharing radios, deteriorated rapidly. Cellular phones offered no alternative, as hundreds of journalists rushed to their phones and overloaded the phone network. Within the first hour of the operation, the Jefferson County dispatch center lost access to the local command post because of jammed radio links. Steve Davis, public information officer of the Jefferson County Sheriff's Office, later commented that "[r]adios and cell[ular] phones and everything else were absolutely useless, as they were so overwhelmed with the amount of traffic in the air."[2]

nologies of different responders must not interfere with one another's operation, and the systems into which these technologies are integrated must be designed so that they complement the users rather than distract them from their missions.[47]

One important area of research is in defining low-level communication protocols and developing generic technology that can facilitate interconnection and interoperation of diverse information resources.[48] For example, software-programmable waveforms can (in principle) allow a single radio to interoperate with a variety of wireless communications protocols; research in this area has gone forward, but further research is needed.[49] A second example is an architecture for communications, per-

---

[47]CSTB, NRC, 1999, *Information Technology Research for Crisis Management*, pp. 50, 84; *Computing and Communications in the Extreme*, p. 33.

[48]CSTB, NRC, 1999, *Information Technology Research for Crisis Management*, p. 85.

[49]Computer Science and Telecommunications Board, National Research Council. 1997. *The Evolution of Untethered Communications*. National Academy Press, Washington, D.C.

> ### World Trade Center
>
> On the morning of September 11, 2001, a few minutes after the South Tower of the World Trade Center had collapsed, police helicopters inspecting the North Tower reported that "about 15 floors down from the top, it looks like it's glowing red" and that collapse was "inevitable." A few seconds later, another pilot reported, "I don't think this has too much longer to go. I would evacuate all people within the area of that second building." Transmitted 21 minutes before the North Tower fell, these messages were relayed to police officers, most of whom managed to escape. However, most firefighters in the North Tower never heard those warnings, nor did they hear earlier orders to get out, and the *New York Times* estimated that at least 121 firefighters were in the tower when it collapsed. The Fire Department radio system had proven unreliable that morning, and it was not linked to the police radio system. Moreover, the police and fire commanders on the scene had only minimal interactions to coordinate strategy or to share information about building conditions.
>
> ---
>
> SOURCE (Columbine): Adapted from Viktor Mayer-Schönberger. 2002. *Emergency Communications: The Quest for Interoperability in the United States and Europe*. BCSIA Discussion Paper 2002-7. John F. Kennedy School of Government, Harvard University, Cambridge, Mass., March. (World Trade Center): Adapted from Jim Dwyer, Kevin Flynn, and Ford Fessenden. 2002. "9/11 Exposed Deadly Flaws in Rescue Plan." *New York Times*, July 7. Available online at <http://www.nytimes.com/2002/07/07/nyregion/07EMER.html?pagewanted=1>.
>
> [1] The description and analysis of the Columbine High School incident is based on two John F. Kennedy School of Government cases: "The Shootings at Columbine High School: Responding to a New Kind of Terrorism," Case No. C16-01-1612.0, and "The Shootings at Columbine High School: Responding to a New Kind of Terrorism Sequel," Case No. C16-01-1612.1.
>
> [2] Id, at 16.

haps for selected mission areas, that translates agency-specific information into formats and semantics compatible with a global system.[50] More generally, "translation" technology can be developed to facilitate interoperable communications among emergency responders (e.g., technology that facilitates interoperability between disparate communications or database systems).

Note also that new technical approaches are not the only option for helping to facilitate interoperable crisis communications. For example, it is likely that some portion of the public networks would survive any disaster; emergency-response agencies could use that portion to facilitate

---

[50] CSTB, NRC, 1996, *Computing and Communications in the Extreme*, p. 17; CSTB, NRC, 1999, *Information Technology Research for Crisis Management*. An additional discussion of mission slices and working the semantic interoperability problem appears in CSTB, NRC, 1999, *Realizing the Potential of C4I*.

BOX 3.6 Why Interoperability Is Difficult

Interoperability is difficult in an emergency-responder context for many reasons. Among them are the following:

- *A lack of common operating frequencies.* For historical reasons, various public service agencies, from law enforcement to firefighters to emergency medical services have traditionally used different frequency bands for their radio communications.[1] Thus, different agencies using different radio systems have generally resorted to sharing or carrying radios from other agencies to facilitate interoperability—an obviously clumsy solution.
- *Inadequacies of analog radio technology used for emergency commmunications.* Today's emergency communications are largely based on analog technology. Though analog technologies have some advantages for emergency responders, such as greater robustness in the presence of clutter and freedom from line-of-sight restrictions (see Box 3.4), they are generally unable to handle the enormous traffic volume that characterizes severe incidents.
- *Independence of the procuring agencies.* As a general rule, response agencies acquire IT systems independently, with little coordination of objectives or requirements, whereas optimizing overall system performance requires a full understanding of the trade-offs entailed by different choices. Individual agencies, especially those that seldom interact with other units, are strongly motivated to solve their own pressing problems, even if doing so makes it harder for them to interact with other units.
- *Inability to anticipate all relevant scenarios for use.* It is difficult to anticipate in detail how information systems will be used—a difficulty that is multiplied in an uncertain environment. For example, the responders to a large-scale terrorist attack are likely to include agencies that have not worked together in the past. Achieving flexibility in such a situation depends heavily on building in a sufficient degree of interoperability, and yet if a given scenario has not been anticipated, the would-be responders may have no particular idea about who they will need to work with.
- *Inclusion of legacy systems.* Legacy systems are in-place systems that are relatively old and were not designed to be easily integrated with current and future information systems, but which remain absolutely essential to the functioning of an organization. Furthermore, they often represent significant investment, so replacing them with new, more interoperable systems is not a near-term option. Today, these legacy systems include the vast installed base of conventional analog radio equipment used by most emergency-response agencies.

SOURCES: For the first bullet item above, Viktor Mayer-Schönberger, 2002, *Emergency Communications: The Quest for Interoperability in the United States and Europe*, BCSIA Discussion Paper 2002-7, John F. Kennedy School of Government, Harvard University, Cambridge, Mass., March; for second item, Box 3.4 in this report; for third through fifth items, Computer Science and Telecommunications Board, National Research Council, 1999, *Realizing the Potential of C4I*, National Academy Press, Washington D.C.

[1] The Federal Communications Commission allocates 13 discrete portions of spectrum for public safety operations but does not specify the types of public safety agencies that are required to use each portion; see Public Safety Wireless Network,1999, *Spectrum Issues and Analysis Report*.

interoperability if there were mechanisms for giving them first priority for such use. A second option would be to allocate dedicated spectrum bands for emergency responders and to require by law that they use those frequencies. (This option is likely to be undesirable under most circumstances, because for routine work a high degree of interoperability is not required, and sharing frequencies might well interfere with such work.) A third option would be to mandate frequency and waveform standards for emergency responders so that they are interoperable. These policy options are not mutually exclusive, and all might benefit from technologies described above.

3.2.2 Emergency Deployment of Communications Capacity

In an emergency, extraordinary demands on communications capacity emerge. A disaster (or an attack on the communications of emergency responders in conjunction with a physical attack) is likely to destroy some but not all of the communications infrastructure in a given area, leaving some residual capability. Meanwhile, the disaster provokes greater demands for communication from the general public. The result is often a denial-of-service condition for all, including emergency communications services. The absence of telephone dial tone in a disaster area is common because of increased demands.[51] Even under high-traffic but nonemergency situations, cell-phone networks are sometimes unable to handle the volume of users in a given cell because of statistical fluctuations in call volume, leaving some users without the ability to connect to the network. Nor is the Internet immune to such problems—congestion of shared Internet links, including both last-mile and aggregated feeder links, can cause greatly slowed traffic to occur on facilities that are still operational in a disaster area. Note also that today, the Internet provides modes of communication—e-mail and Web-based information services—that are different from that provided by voice communications.

Given the vastly increased demands for wireless communications that accompany any disaster, efficient spectrum management is essential. One important issue here is the fact that emergency bands need to be kept clear of nonemergency transmissions. However, certain types of nonstandardized but sometimes-deployed higher-speed DSL (digital subscriber line) telecommunications equipment are known to radiate in these bands at unacceptable levels of power, and such equipment can interfere

---

[51]CSTB, NRC, 1996, *Computing and Communications in the Extreme*, p. 17.

with all emergency transmissions and receptions within 50 yards of the offending telephone lines.[52]

In addition, it may be possible to improve the use of available spectrum in a disaster area through the use of processing and portable hardware (network equipment or antennas carried by truck into area) that would nominally exceed cost constraints for general deployment. Such equipment ranges from truck-portable cellular stations to antennas usable for receiving and rebroadcasting degraded emergency-band communications. Vector signal processing of signals from multiple antennas deployed rapidly and ubiquitously can increase available spectrum by an order of magnitude or more. Fast reconnection to high-performance equipment can also greatly increase the amount of information flow to and from affected areas.

Research is needed on using residual (and likely saturated) capacity more effectively, deploying additional ("surge") capacity,[53] and performing the trade-offs among different alternatives. One problem in this area is the management of traffic congestion and the development of priority overrides for emergency usage (and prevention of the abuse of such authority).[54] For example, the capacity of DSL links could be significantly increased through coordination across neighboring telephone lines (bandwidth increases of an order of magnitude are sometimes possible). Under normal circumstances, signals transmitted by other users result in cross-

---

[52] The specific type of equipment is known as "single-carrier VDSL (very high data rate digital subscriber line)." It is sold by several vendors, despite some local exchange carriers reporting that it interferes with the reception of any amateur radio receiver (which is the same type used in emergencies) within at least 30 meters of an aerial telephone line. At least one local exchange carrier is known to have deployed it in a VDSL trial in Phoenix, Arizona, despite its violation of the radio interference requirements. (John Cioffi, Stanford University, personal communication, July 9, 2002.)

[53] CSTB, NRC, 1999, *Information Technology Research for Crisis Management*, p. 83.

[54] CSTB, NRC, 1999, *Information Technology Research for Crisis Management*, pp. 39, 52; CSTB, NRC, 1996, *Computing and Communications in the Extreme*, p. 20. In addition, the White House's National Communications System (NCS) office has moved to implement a wireless priority service that facilitates emergency-recovery operations for the government and local emergency-service providers. This service will be implemented in phases, with an immediate solution available in early 2002 in selected metropolitan areas and a nationwide solution (yet to be developed) scheduled for late 2003. Further work after 2003 will concentrate on the development and implementation of "third-generation" technologies that enable high-speed wireless data services. See Convergence Working Group. 2002. *Report on the Impact of Network Convergence on NS/EP Telecommunications: Findings and Recommendations*. February.

talk that limits available bandwidth, but in an emergency those interests could be reprioritized.[55]

A second problem is that optimization algorithms for communications traffic that are appropriate in normal times may have to be altered during emergencies. For example, the destruction of physical facilities such as repeaters and the massive presence of debris could result in an impaired environment for radio transmissions. The rapid deployment of processors optimized to find weak signals in a suddenly noisier environment could do much to facilitate emergency communications.

Research is also needed to develop self-adaptive networks that can reconfigure themselves in response to damages and changes in demand, and that can degrade gracefully.[56] For example, in a congested environment, programmed fallback to less data-intensive applications (e.g., voice rather than video, text messaging rather than voice) may provide a minimal communications facility. Even today, many cellular networks allow the passing of text messages. Also, both public and private elements of communications infrastructure could be tapped to provide connectivity in a crisis, as happened in New York City on September 11.[57]

3.2.3 Security of Rapidly Deployed Ad Hoc Networks

The management of communications networks poses unique problems in the crowded emergency disaster zone. Security must be established rapidly from the outset, as terrorists might try to infiltrate as emergency agencies responded.[58] It is also necessary to determine a means for temporarily suspending people's access to facilities, communications, and data without impeding the ability of those with legitimate need to use them. But this suspension process has to be done rapidly, given that minutes and seconds matter in severe emergencies. Note also that large deployments may well pose security problems that are more severe than those posed by smaller deployments.

---

[55] Actions that disadvantage commercial consumers in times of emergency have some precedent. Of particular relevance is the Civil Reserve Air Fleet program of the DOD, under which priority use of commercial airliners is given to the Defense Department in times of national emergency. (Because ordinary airliners lack the strong floors needed to support heavy military equipment, the floors are strengthened at government expense, and the government pays the incremental fuel costs of carrying that excess weight.)

[56] CSTB, NRC, 1999, *Information Technology Research for Crisis Management*, p. 39.

[57] CSTB, NRC, 1999, *Information Technology Research for Crisis Management*, p. 39.

[58] CSTB, NRC, 1996, *Computing and Communications in the Extreme*, p. 24. *Embedded, Everywhere* (CSTB, NRC, 2001) also discusses security issues associated with ad hoc networks, though in the context of sensor networks rather than communications networks for human beings.

Research is therefore needed on the special security needs of wireless networks that are deployed rapidly and in an ad hoc manner. (For example, ad hoc networks are not likely to have a single system administrator who can take responsibility for allocating user IDs.)

3.2.4 Information Management and Decision-Support Tools

Emergency responders have multiple information needs that technology can support.[59] For example, a multimedia terminal for a firefighter could have many uses: [60] accessing maps and planning routes to a fire, examining building blueprints for tactical planning, accessing databases locating local fire hydrants and nearby fire hazards such as chemical plants, communicating with and displaying the locations of other fire and rescue teams, and providing a location signal to a central headquarters so that a firefighting team can be tracked for broader operational planning. Note that only some of this data can be pre-stored on the terminal, especially because some data may have to be updated during an operation, so real-time access to appropriate data sources is necessary. Moreover, these different applications require different data rates and entail different trade-offs between latency and freedom from transmission errors. Voice communications, for example, must have low latency (i.e., be real time) but can tolerate noisy signals, whereas waiting a few seconds to receive a map or blueprint will generally be tolerable, though errors may make it unusable. Some applications, such as voice conversation, require symmetrical bandwidth; others, such as data access and location signaling, are primarily one-way (the former toward the mobile device, the latter away from it).

Another issue arises because disasters create huge volumes of information to be handled. A large volume of voice and data traffic will be transmitted and received on handheld radios, phones, digital devices, and portable computers. Large amounts of additional information will be available from various sources including the World Wide Web, building blueprints, and interviews with eyewitnesses. However, sorting out reliable and useful information from this vast array of sources is inevitably difficult, and many crisis responders have adopted a rule of thumb about such information: one-third of the information is accurate, one-third is

---

[59]This discussion is taken from CSTB, NRC, 1996, *Computing and Communications in the Extreme*, p. 59.

[60]Randy H. Katz. 1995. "Adaptation and Mobility in Wireless Information Systems." Unpublished paper, available online at <http://daedalus.cs.berkeley.edu>. August 18.

wrong, and one-third might be either right or wrong.[61] Such inaccuracies further compound the difficulties of integrating such data.

Thus, emergency-response managers need tools that can draw on information from diverse and unanticipated sources to assist them in evaluating, filtering, and integrating information and in making decisions based on incomplete knowledge of conditions, capabilities, and needs.[62] Such tools would interpret information about the quality and reliability of varied inputs and assist the user in taking these variations into account. These tools would differ from, but could build upon, traditional decision-support tools such as knowledge-based systems, which operate using rules associated with previously examined problems. Because many of the problems raised by crisis management are not known ahead of time, more general techniques are needed. These might include the development of new representations of the quality of inputs (such as metadata about those inputs regarding qualities such as reliability and uncertainty in the data), data fusion, models of information representation and integration (e.g., integrating pre-existing disaster response plans with decision-support tools), rapidly reconfigurable knowledge-based systems, and new techniques for filtering and presenting information to users and for quickly prioritizing tasks. Section 3.3 addresses some of these issues in a broader context.

3.2.5 Communications with the Public During an Emergency

In a crisis, channels to provide information to the public are clearly needed. Radio, television, and often the Web provide crisis information today, but such information is usually generic and not necessarily helpful to people in specific areas or with specific needs. Information tailored to specific individuals or locations through location-based services (and perhaps simple response systems to ascertain the location of the injured and to identify critical time-urgent needs) could be put to more effective use.

Research is needed to identify appropriate mechanisms—new technologies such as "call by location" and zoned alert broadcasts—for tailoring information to specific locations or individuals.[63] To enable effective interaction with individual users, ubiquitous and low-cost access is re-

---

[61]CSTB, NRC, 1999, *Information Technology Research for Crisis Management*, Appendix B.

[62]CSTB, NRC, 1996, *Computing and Communications in the Extreme*, National Academy Press, Washington, D.C., p. 104; CSTB, NRC, 1999, *Information Technology Research for Crisis Management*, pp. 32-33.

[63]CSTB, NRC, 1999, *Information Technology Research for Crisis Management*, p. 35.

quired.[64] In addition, such systems should be highly robust against spoofing (entry by an intruder masquerading as a trusted party) so that only authorized parties can use the systems to send out information.

For example, the current cell-phone system does not directly support these functions, but it might be possible to modify and exploit it to provide "reverse 911" service,[65] that is, a one-way channel to people affected by the crisis that provides a flow of relevant information and guidance. Such mechanisms would probably have to be locally sufficient. That is, the disaster might spare the local cell site—or a temporary cell site could be deployed along with wireless alternatives[66]—but access to central services might not be possible.

Finally, providing information to those located outside the immediate emergency area provides important psychological comfort and helps to mitigate a disaster's consequences. (For example, in the immediate aftermath of the September 11 attacks, "I'm alive" Internet bulletin boards sprang up spontaneously.) Research is needed for establishing more effective means of achieving this objective—especially in updating the status of affected people—while compromising the local communications infrastructure to a minimal degree.

3.2.6 Emergency Sensor Deployment

During an emergency, responders need information about physical on-the-ground conditions that is sufficiently fine-grained and accurate to be useful. However, existing sensor networks may be disrupted or destroyed to some extent—and it is important to be able to recognize an attack on a sensor network and to monitor the state of the network's health so that emergency responders have an idea of the extent to which they can continue relying on the information provided by the network. When the surviving sensor capability cannot provide adequate information, the deployment of additional sensors is likely to be necessary. Depending on the nature of the emergency, sensor capacity would be needed to identify and track the spread of nuclear, chemical, or biological contaminants, characterize and track vehicular traffic, locate survivors (e.g., through heat emanations, sounds, or smells), and find pathways through debris and rubble. Developing robust sensors for these capabilities is one major challenge; developing architectural concepts for how to deploy them and integrate the resulting information is another.[67]

---

[64]CSTB, NRC, 1999, *Information Technology Research for Crisis Management*, p. 40.
[65]CSTB, NRC, 1999, *Information Technology Research for Crisis Management*, p. 35.
[66]CSTB, NRC, 1996, *Computing and Communications in the Extreme*, p. 18.
[67]CSTB, NRC, 2001, *Embedded, Everywhere*.

3.2.7 Precise Location Identification

When extensive physical damage to a structure or an area occurs, determining the location of physical structures and of people is a major problem. One reason is that many reference points for observers on the ground or in the structure may become unavailable: debris, airborne contaminants such as smoke and dust, and perhaps simply the lack of illumination can reduce visibility, and street signs may disappear. Also, the physical environment changes to some extent as a result of physical damage: fires destroy rooms in buildings, one floor collapses into another, a crater exists where there used to be three buildings.

Thus, technology must be available to help establish reference points when existing ones are destroyed or inadequate. In an open physical environment, location finders based on the Global Positioning System (GPS) serve a useful role, but airborne contaminants, equipment damage, and line-of-sight restrictions could adversely affect the reliability of GPS, and research is needed on ways to determine location in a disaster environment. Embedded location sensors and sensor networks, either already in place or deployed in response to an incident, are likely to be valuable information sources.[68]

Responders and victims also require rapid access to accurate locational databases. One such source is pre-existing building blueprints and diagrams.[69] Where these do not exist, or where the structures themselves have suffered significant physical damage, on-the-spot mapping will inevitably be needed. Thus, research is needed to develop digital floor plans and maps of other physical infrastructure.[70] The resulting data could be stored in geospatial information systems (GIS), which would allow responders to focus on the high-probability locations of missing people (such as lunchrooms), and avoid dangerous searches of low-probability locations (such as storage areas).[71] Research is needed in wearable computers for search-and-rescue operations[72] so that responders could update the GIS in real time as they discover victims and encounter infra-

---

[68]CSTB, NRC, 1996, *Computing and Communications in the Extreme*, pp. 24, 25; CSTB, NRC, 2001, *Embedded, Everywhere*.

[69]J. Hightower and G. Boriello. 2001. "Location Systems for Ubiquitous Computing." *IEEE Computer*, Vol. 33 (8).

[70]As one example, consider that a firm that installs fiber-optic cables in a city's sewers is capable of mapping those sewers as well using a sewer-crawling robot that lays cable and tracks its position.

[71]CSTB, NRC, 1996, *Computing and Communications in the Extreme*, p. 14.

[72]CSTB, NRC, 1999, *Information Technology Research for Crisis Management*, p. 38.

structural damage. Another research area is in "map ants"[73]—distributed, self-organizing robots deployed in a disaster area to sense movement or body heat, for instance. It may also be possible to develop technology to generate the data for accurate maps of a debris-strewn disaster location.

Finally, keeping track of emergency responders' positions within a disaster area is an essential element of managing emergency response. Technology (similar to E-911 for cell phones) to monitor the progress of these individuals automatically is not yet available on a broad scale.

3.2.8 Mapping the Physical Aspects of the Telecommunications Infrastructure

As noted in Section 2.2.2, the telecommunications infrastructure is for the most part densely connected; hence physical attack is unlikely to disrupt it extensively for long periods of time. However, the physical infrastructure of telecommunications (and the Internet) does not appear to be well understood (that is, immediate knowledge of where various circuits are located is in many cases available only in a disaggregated form distributed throughout a myriad of company databases), and there may well exist critical nodes whose destruction would have disproportionate impact. (On the other hand, knowing where these critical nodes are is difficult for both network operators and terrorists.) Thus, an important priority is to develop tools to facilitate the physical mapping of network topology, and to begin that mapping with the tools that are available today. This is particularly important for the many converged networks over which both voice and data are carried.

3.2.9 Characterizing the Functionality of Regional Networks for Emergency Responders

In order to develop mechanisms for coordinating emergency-response activities, it is necessary to understand what the various communications and computer networks of emergency responders in a given region are supposed to do. For example, managers from different agencies often speak different "languages" in describing their needs, capabilities, and operational priorities; a common conceptual framework for these purposes would be enormously helpful for the coordination of planning ac-

---

[73]A study in progress by the Computer Science and Telecommunications Board on the intersections between geospatial information and information technology discusses these self-organizing robots. See <http://www.cstb.org> for more information on this study.

tivities, yet one is not available.[74] Sharing of information among the various providers of critical infrastructure and emergency-response agencies, even about common tasks and processes, has been a rather uncommon activity in the past.

An additional and challenging problem is that severe emergencies often change what different agencies are expected to do and what priorities their duties have. For example, existing plans are often overridden by circumstances (e.g., in the aftermath of the September 11 attacks, phone services to the financial district in Wall Street were given priority; this eventuality had not been anticipated in prior emergency planning). Moreover, even if existing plans do not require alteration, emergency agencies must know and be able to execute on what those plans contain. (For example, the notebook containing personnel assignments in a certain type of emergency must be available and easily found.) For this reason, drills and exercises are necessary, and they must involve enough personnel that in a real emergency, some people with drill experience will be participating in actual response operations.

3.3 INFORMATION FUSION

As discussed below, information fusion promises to play a central role in the future prevention, detection, attribution, and remediation of terrorist acts. Information fusion is defined as the acquisition of data from many sources, the integration of these data into usable and accessible forms, and their interpretation. Such integrated data can be particularly valuable for decision makers in law enforcement, the intelligence community, emergency-response units, and other organizations combating terrorism. Information fusion gains power and relevance for the counterterrorist mission because computer technology enables large volumes of information to be processed in short times. (Box 3.7 provides elaborated scenarios for ways in which information-fusion techniques could be useful.)

- *Prevention.* Security checkpoints have become more important, and more tedious, than ever at airports, public buildings, sporting venues, and national borders. The efficiency and effectiveness of checkpoints could be significantly improved by creating information-fusion tools to

---

[74]Hank Christen, Paul Maniscalco, Alan Vickery, and Frances Winslow. 2001. "An Overview of Incident Management." *Perspectives on Preparedness*, No. 4, September. Available online at <http://ksgnotes1.harvard.edu/BCSIA/Library.nsf/pubs/POP4>.

BOX 3.7 Scenarios for Automated Evidence Combination

Intelligence Analysis

A video monitoring program detects an unknown person speaking at a meeting with a known terrorist in an Arabic television broadcast. This information is brought to the attention of the intelligence analyst who set up the program to scan TV for videoclips that mention individuals known to associate with a particular terrorist group. The analyst decides to find out as much as possible about this new unknown person. First, a check is run on video archives using face detection/recognition software to see if this person's face had appeared before but was missed. Second, the recording of the person's speech from the videoclip is used to match against stored archives of conversations. Third, the person's name, mentioned by someone in the videoclip, is used to search information resources that are both internal (across intelligence agencies) and external (such as the Web and multilingual news archives). The search is focused on text that mentions this name in the context of terrorism or in connection with any of the activities of the known terrorist. The material that is found is summarized in a variety of ways to aid analysis, such as in a time line with geographic locations.

The result of this search indicates that the unknown person works for an aid organization that operates in a number of countries, and that the person has recently visited the United States. A request for appropriate legal authority results in a targeted search of the hotel, air travel, car rental, and telephone transactions associated with this trip. This search reveals connections to other people that the FBI has had under surveillance in the Washington, D.C., area, including through court-approved wiretaps. The automatic analysis of phone calls between the unknown person and those already under surveillance also indicates an unusual pattern of language including phrases like "wedding party" and "special delivery" in Arabic. These phrases are subsequently linked to overseas intelligence intercepts. On the basis of this information, the analyst concludes that there may be something being planned for a specific time period and thus initiates a number of additional automatic monitoring and alerting tools focused on these people and these language patterns.

Airport Security Screening

At airport security checkpoints, every carry-on bag is x-rayed, and some percentage (say, 15 percent) of these bags receives a manual examination searching for items that could be used as weapons. Though this is similar to search procedures in operation today, this scenario also entails checkpoint operators supported by a network of computers in this airport that captures image, sound, x-ray, and other sensor information; analyzes these data using computer perception software; shares the data across a national computer network that links every airport in the United States; assists human security agents; can be instructed to monitor for certain people or events as needed; and learns from experience what is normal and abnormal within each of the nation's airports.

First, the x-ray machines that capture images of each bag are connected to a computer network. As each bag goes through the machine, a computer vision system analyzes it to attempt to detect suspicious items, such as sharp objects and guns, and alerts the operator if anything is detected, highlighting it on the x-ray screen.

This computer vision system is not infallible, but then neither is the human baggage screener, and the two working together notice different things, so that together they are more effective than either alone (if they make the same number of errors, but independently, their combined error rate is halved).

Second, as the manual checks are performed on 15 percent of the bags, the computer is informed about which bags are checked and whether anything was found. This information provides training data to the computer, which is tied back to the x-ray image of the bag, enabling the vision system over time to improve its ability to detect suspicious items from the x-rays. Standard machine-learning algorithms are used for this purpose, not unlike those used in face recognition. Of course, the human operator will learn on the job in this way as well, but there is one major difference: the human operator may inspect hundreds or a few thousand bags in a week, whereas the computer network is capturing data from every security station in every airport around the world. Thus, the computer has thousands of times more training experience to learn from in a single day than any human operator could accumulate in a lifetime, leading it to much more refined capabilities.

Because of these extensive training experiences, the computer network is able to learn things that a single operator never could. For example, it will learn which new models of suitcases, briefcases, and purses have become common, and it will therefore be able to spot bags that may be custom-built by a terrorist who has designed luggage for a special purpose. It will learn what are typical false alarms raised by each model as it goes through security (e.g., the metal clasp on a particular purse might often be mistaken as a blade by those unfamiliar with that luggage model). The computer system advises the security operator about which 15 percent of the bags should be inspected, decreasing the chance that the operator misses an item or a person who should not go unnoticed. Note that the human operator is still in charge but is greatly aided by the vast experience base gained by the self-improving network of computer sensors and image analyzers.

Third, the computer network can detect patterns involving multiple people entering through multiple security stations, or at multiple times. If there are several people taking the same flight, each of whom has "accidentally" brought a pocket knife, this would be detected by the system, even though the individual operators at the different security posts would likely be unaware of this global pattern. Global patterns could also be detected instantly across multiple airports (e.g., single passengers boarding with similar "accidental" pocketknives in Boston, San Francisco, Washington, D.C., and New York City).

Fourth, the security checkpoint operator will be informed by the system of events that occurred more globally throughout the airport and in airports with connecting flights, as captured by security cameras and other sensors throughout the airport. For example, if two people arrived at the airport together (as indicated by the cameras elsewhere in the airport) but get in different lines going through security, the operator might be informed of this somewhat unusual behavior. While no single clue such as this is clear-cut evidence of a dangerous situation, the security station operators are more effective because the network of computer sensors and analysis provides them a more global and more informed picture of what is going on in the airport, enabling them to focus their attention on the most significant risks.

Finally, the information garnered, learned, and transmitted across this network of airport monitoring systems is used to help train individual security operators and to analyze how past security breaches were successful.

support the checkpoint operator in real time. For example, future airport-security stations could integrate data received from multiple airports to provide a more comprehensive view of each passenger's luggage and activities on connecting flights. The stations could use data-mining methods to learn which luggage items most warrant hand-inspection, and they could capture data from a variety of biometric sensors to help verify the identities of individuals and to search for known suspects.

- *Detection and attribution.* Intelligence agencies are routinely involved in information fusion as they attempt to track suspected terrorists and their activities. One of their primary problems is that of managing a flood of data. There are well-known examples in which planned terrorist activity went undetected despite the fact that evidence was available to spot it, because the evidence was just one needle in a huge haystack. Future intelligence and law-enforcement activities could therefore benefit enormously from advances in the automated interpretation of text, image, video, sensor, and other kinds of unstructured data. Automated interpretation would enable the computer to sort efficiently through massive quantities of data to bring the relevant evidence (likely combined from various sources) to the attention of an analyst.

- *Remediation (response).* Early response to biological attacks could be supported by collecting and analyzing real-time data, such as ambulance calls, admissions to hospital emergency rooms and veterinary offices, or purchases of nonprescription drugs in grocery stores to identify spreading disease. Prototype systems are already under development (Box 3.8). If anomalous patterns emerge that may signify an outbreak of some new pathogen, system administrators can quickly alert health officials.

Many other opportunities exist for such computer-aided evidence-based decision making. For example, the monitoring of activity on computer networks might flag potential attempts to break through a firewall; or, sensor networks attached to public buildings might flag patterns of activity within the building that suggest suspicious behavior. In such cases, an unaided decision maker might have difficulty detecting subtle patterns, because the data are voluminous and derive from a variety of sources.

As a general proposition, the development of tools that provide human analysts with assistance in doing their jobs has a higher payoff (at least in the short to medium term) than do tools that perform most or all of the analyst's job. The former approach places a greater emphasis on using technology to quickly sift large volumes of data in order to flag potentially interesting items for human attention, whereas the latter ap-

BOX 3.8 Examples of Information Fusion for Detection of Bioterrorism Attacks

• The Biomedical Security Institute's Real-time Outbreak and Disease Surveillance System monitors about 1,200 patient visits per day in 17 western Pennsylvania hospitals and recognizes patterns of infectious disease.[1] The system looks for sudden and frequent outbreaks of cases involving flu-like symptoms, respiratory illnesses, diarrhea, and paralysis that might indicate a bioterror attack. A sudden peak in a certain type of symptom may be the result of a biological agent. Emergency rooms and hospitals provide patient information such as symptoms, age, gender, address, and test results directly to the system.[2] Public health doctors are automatically notified if a pattern develops, and they in turn notify the proper authorities.

• The Centers for Disease Control and Prevention supports a number of enhanced surveillance projects (ESPs) that monitor certain hospital emergency departments to establish baseline rates of various clinical syndromes.[3] Anomaly detection models identify significant departures from these baseline rates that are called to the attention of state and local health departments for confirmation and appropriate follow-up. ESP has been tested at the 1999 World Trade Organization Ministerial in Seattle, the 2000 Republican and Democratic National Conventions held in Philadelphia and Los Angeles, respectively, and the Super Bowl/Gasparilla Festival in Tampa, Florida. A similar effort is the Lightweight Epidemiology Advanced Detection and Emergency Response System (LEADERS), which provides a set of tools originally developed for the military to provide real-time analysis of medical data entered at health care delivery points (e.g., hospitals). These data are examined for spatial correlations so that areas of potential disease outbreak can be rapidly identified.

• A system in Washington, D.C., called ESSENCE II (ESSENCE stands for Electronic Surveillance System for Early Notification of Community-based Epidemics) is an operational prototype for the National Capital Region that integrates information on nonmedical information such as employee absenteeism and over-the-counter drug sales with information available from doctors' visits, diagnostic laboratories, hospital emergency rooms, 911 and poison control calls, and so on. These data are acquired on a near-real-time basis, and are examined to search for temporal and/or spatial anomalies that might indicate an early warning of a bioterrorist attack. Nonmedical information is valuable in this context because it is likely to reflect changes in public health status before affected individuals seek medical care on a scale large enough to register a possible attack.[4]

---

[1] Bruce Gerson. 2002. "President Bush Praises Carnegie Mellon, Pitt Collaboration to Fight Bioterrorism." *Carnegie Mellon News*. February 13. Available online at <http://www.cmu.edu/cmnews/extra/020213_gbush.html>.

[2] James O'Toole et al. 2002. "Bush Here Today to Highlight Spending on Terrorism." *Pittsburgh Post-Gazette*, February 5. Available online at <http://www.post-gazette.com/regionstate/20020205visit0205p2.asp>.

[3] See <http://ndms.umbc.edu/conference2001/2001con20/Treadwe.htm>.

[4] For more information on ESSENCE II, see the Web site at <http://www.nyam.org/events/syndromicconference/agenda.shtml>.

proach relies on computers themselves to make high-level inferences in the absence of human involvement and judgment.

A final dimension of information fusion is nontechnical. That is, disparate institutional missions may well impede the sharing of information. Underlying successful information-fusion efforts is a *desire* to share information—and it is impossible to fuse information belonging to two agencies if those two agencies do not communicate with each other. Establishing the desire to communicate among all levels at which relevant information could be shared might have a larger impact than the fusion that may occur due to advances in technology. For example, hospitals generally see little benefit and many risks in sharing patient information with other hospitals, even if it may facilitate the care of some patients that use both facilities or improve epidemiologic monitoring. (Note also that achieving a desire for different facilities or agencies to communicate with each other may take more than moral persuasion. This point is addressed in greater detail in Sections 3.6.3 and 3.6.4.)

3.3.1 Data Mining

Data mining is a technology for analyzing historical and current online data to support informed decision making. It has grown quickly in importance in the commercial world over the past decade, because of the increasing volume of online data, advances in statistical machine-learning algorithms for automatically analyzing these data, and improved networking that makes it feasible to integrate data from disparate sources.

The technical core of data mining is the ability to automatically learn general patterns from a large volume of specific examples. For example, given a set of known fraudulent and nonfraudulent credit-card transactions or insurance claims, the computer system may learn general patterns that can be used to flag future cases of possible fraud. Other applications are in assessing mortality risk for medical patients (by learning from historical patient data) and in predicting which individuals are most likely to make certain purchases (by analyzing other individuals' past purchasing data). Decision-tree learning, neural-network learning, Bayesian-network learning, and logistic-regression-and-support vector machines are among the most widely used statistical machine-learning algorithms. Dozens of companies now offer commercial implementations, which are integrated into database and data-warehousing facilities.

The majority of these commercial data-mining applications involve well-structured data. However, current commercially available technology does not work well with data that are a combination of text, image, video, and sensor information (that is, data in "nonstructured" formats).

Moreover, it is largely unable to incorporate the knowledge of human experts into the data-mining process. Despite the significant value of current machine-learning algorithms, there is also a need to develop more accurate learning algorithms for many classes of problems.

New research is needed to develop data-mining algorithms capable of learning from data in both structured and nonstructured formats. And, whereas current commercial systems are very data-intensive, research is needed on methods for learning when data are scarce (e.g., when there are only a few known examples of some kinds of terrorist activity) by incorporating the knowledge of human experts alongside the statistical analysis of the data. Another research area is that of better mixed-initiative methods that allow the user to visualize the data and direct the data analysis.

3.3.2 Data Interoperability

An inherent problem of information fusion is that of data interoperability—the difficulty of merging data from multiple databases, multiple sources, and multiple media. Often such sources will be distributed over different jurisdictions or organizations, each with different data definitions. New research is needed for normalizing and combining data collected from multiple sources, such as the combination of different sets of time-series data (e.g., with different sampling rates, clocks, and time zones), or collected with different data schemas (e.g., one personnel database may use the variable "JobTitle," while another uses "Employee Type"). Note that data interoperability is also an essential element of organizational and operational interoperability.

3.3.3 Natural Language Technologies

In the past decade, the area of language technologies has developed a wide variety of tools to deal with very large volumes of text and speech. The most obvious commercial examples are the Web search engines and speech recognition systems that incorporate technology developed with funding from the Defense Advanced Research Projects Agency (DARPA) and the National Science Foundation (NSF). Other important technologies include information extraction (e.g., extracting the names of people, places, or organizations mentioned within a document), cross-lingual retrieval (e.g., does an e-mail message written in Arabic involve discussion of a chemical weapon?), machine translation, summarization, categorization, filtering (monitoring streams of data), and link detection (finding connections). Most of these approaches are based on statistical models of language and machine-learning algorithms.

A great deal of online information, in the form of text such as e-mail, news articles, memos, and the Web, is of potential importance for intelligence applications. Research is needed on methods for accurately extracting from text certain structured information, such as descriptions of events (e.g., the date, type of event, actors, and roles.) Research is needed to handle multiple languages, including automatic translation, cross-lingual information retrieval, and rapid acquisition of new languages. Other important areas of future research are link detection (related to the normalization problem mentioned above) and advanced question answering.

3.3.4 Image and Video Processing

The technologies for image and video processing tend to be domain-specific and often combine information from multiple modes. For example, several companies are beginning to offer image-recognition software for face recognition and automatic classification of medical and other types of images. Commercially available video indexing-and-retrieval software improves effectiveness by combining techniques of segmentation, face detection, face recognition, key frame extraction, speech recognition, text-caption extraction, and closed-caption indexing. This is a good example of information fusion in which multiple representations of content are combined to reduce the effect of errors coming from any given source.

The major limitation of present language and image technologies is in accuracy and performance: despite significant progress, these need to be considerably improved. This is particularly important for counterterrorist systems where the data may be very noisy (i.e., surrounded by irrelevant information) and sparse.

Work is needed on improved algorithms for image interpretation and speech recognition. Many of these research issues are specific to problems arising in a particular medium—for example, recent progress on face recognition has come primarily from understanding how to extract relevant image features before applying machine-learning methods, though this approach may not be applicable to machine learning in other contexts. However, new research is also needed on perception based on mixed media, such as speech recognition based on sound combined with lip motion.

3.3.5 Evidence Combination

Many of the techniques used to combine information from multiple sources, as in video indexing or metasearch engines, are ad hoc. Current research on more principled methods for reasoning under uncertainty needs to be extended and tested extensively in more demanding applications. This is a key technical problem, with widespread implications for many of the applications mentioned above—for example, how to combine evidence from hospital admissions and from nonprescription drug purchases to detect a possible bioterrorist attack, how to combine evidence from face recognition and voice print to estimate the likely identity of a person, how to combine evidence from multiple sensors in a building to detect anomalous activity, and how to undertake more effective personnel screening.

3.3.6 Interaction and Visualization

All technologies developed for information access and analysis rely on human involvement to specify information problems and to make sense of the retrieved data. Research on human-computer interaction in the areas of query formulation and visualization has the potential for significantly improving the quality and efficiency of intelligence applications. In the area of query formulation, some topics of interest include support for natural-language queries, query triage (deciding which information resource is needed based on the form of the query), and personalization based on the context of previous searches and the current task. A variety of visualization techniques have been developed for large-scale scientific applications, but more research is need on techniques that are effective for visualizing huge amounts of dynamic information derived from unstructured data about people, places and events. Such research is potentially valuable because it takes advantage of the human ability to recognize patterns more easily than automated techniques can.

3.4 PRIVACY AND CONFIDENTIALITY

The essential rationale underlying a science-based approach to countering terrorism is that, absent measures to reduce the intrusiveness and burdens of a counterterrorism regime on individuals, these burdens will become so high that they will be intolerable to the American people over the long run. As pressure mounts for the government to collect and process more information, it becomes increasingly important to address the question of how to minimize the negative impacts on privacy and data confidentiality that may arise with applications of information fusion.

It is axiomatic that information that is not collected can never be abused. For example, consider the possible collection of an individual's HIV status. If the individual has concerns that his or her privacy interests might be compromised by the improper disclosure of HIV status to an employer, then the most effective way to protect such information is not to collect information on HIV status (e.g., by not recording it in medical records or not requiring an HIV test). However, it is also true that information that is not collected can never be used. Thus, if no one ever collects information about the individual's HIV status, that person's medical treatment may be adversely affected.

This dilemma is magnified in a counterterrorism intelligence context. Because terrorists are not clearly identified with any entity (such as a nation-state) whose behavior can be easily studied or analyzed, their individual profiles of behavior and communication are necessarily the focus of an intelligence investigation. Most importantly, it is often not known in advance what specific information must be sought in order to recognize a suspicious pattern, especially as circumstances change. From the perspective of intelligence analysis, the collection rule must be "collect everything in case something might be useful." Such a stance generates obvious conflicts with the strongest pro-privacy rule "Don't collect anything unless you know you need it."

Data mining and information fusion have major privacy implications, and increased efforts by commercial and government entities to correlate data with a specific person negatively impact privacy and data confidentiality. There is no clear and easy way of resolving this tension. Nevertheless, research can help to ameliorate this tension in several ways:

- At a minimum, policy makers need accurate information about the trade-offs between privacy and confidentiality on the one hand and analytical power for counterterrorist purposes on the other under different circumstances of data disclosure. For example, under what circumstances and at what state in an analytical effort are specific identities necessary? What is the impact of increasing less-personalized information? Research is needed to answer such questions.
- Research is needed to mitigate negative aspects of data mining, including research on data-mining algorithms that discover general trends in the data without requiring full disclosure of individual data records. For example, some algorithms work by posing statistical queries to each of a set of databases rather than by gathering every data record into a centralized repository. Others collect aggregate data without requiring full disclosure of individual data records.[75]

---

[75]As a simple example, say that a public health agency is interested in the overall incidence of a particular kind of disease being treated at hospitals in a certain area. One method of obtaining this information is for the agency to ask each hospital for the number

- Research is needed to develop architectures for data storage that protect privacy without compromising analysis. For example, arrangements could be made to use totally separate and distributed data repositories. Using such an approach to information access, an intelligence agency could send a request to other government agencies and "content-providers" asking them if there were any connections between certain individuals. Without revealing the details of the connections, the presence of connections could be made known and then specific legal action could be taken to acquire the data.

Note that institutional concerns about confidentiality can be as strong as individual concerns about privacy. For example, asking hospitals to pass along personally identified patient information to parties not traditionally entitled to such information may well engender institutional resistance that may well not be surmountable by fiat (see the discussion in Section 3.6.3).

For many applications such as badges and access tokens, detailed personal information is not necessary; the only requirements are that the token is recognizable as valid and that it has been issued to the person presenting it. It does not even have to have an individual's name on it. On the other hand, a sufficient aggregation of non-personally-identified information can often be used to identify a person uniquely. Identifying someone as a man of Chinese extraction who has a doctorate in physics, enjoys swing dancing, has an adopted 7-year-old daughter, and lives in upper Northwest Washington, D.C., is probably sufficient to specify a unique individual, even though no particular name is associated with any of these pieces of information. Thus, the mere fact that information is disconnected from personal identifiers is no assurance that an individual cannot be identified if data were aggregated.[76]

---

of cases of that disease and to total them. However, if for some reason a hospital has an incentive to keep that number secret, such a request for data raises privacy concerns. An alternative approach is to ask each hospital to report the sum of the number of cases plus some random number. The agency totals the sums and then asks each hospital to report the random number it used. To obtain the true incidence, it then subtracts out from the total all of the random numbers submitted. In reporting this way, no hospital compromises the true number of cases at that hospital.

[76]It is sometimes surprising how little non-named information is necessary to aggregate in order to re-create an individual's identity. For example, a large number of people can be identified simply by aggregating their date of birth and zip code. See Computer Science and Telecommunications Board, National Research Council. 2000. *IT Research for Federal Statistics*. National Academy Press, Washington D.C.

Another conflict between privacy and intelligence analysis arises because data can be linked. Assuming that various pieces of data have been collected somewhere, intelligence analysis relies on the ability to link data in order to reveal interesting and perhaps significant patterns. Indeed, the very definition of a "pattern of behavior" is one in which data associated with a given individual are grouped so that a trip to Location A can be seen in the context of Transaction B and a phone call with Person C. It is exactly such linkage that gives rise to privacy concerns. Thus, actions taken to increase the interoperability of databases so that meaningful linkages can be created will inevitably raise privacy concerns.

What technical or procedural protections can be developed to guard against inappropriate privacy-compromising linkages but do not also cripple the legitimate search for terrorists? Research in this area is vital. One approach calls for linkages to be uncovered in the relevant computers but not to be revealed except under special circumstances (e.g., when some reasonable cause for suspicion exists[77]). For example, the most recent version of ESSENCE (Box 3.8) uses data that can be rendered anonymous on a sliding scale of anonymity.[78] During normal operation (i.e., when nothing has been detected), sufficiently anonymous data are provided. Once evidence of an attack is encountered, data that are less anonymous and more identifiable, but also more useful for further analysis, are automatically transmitted. Thus, routine compromises of privacy are not entailed, but when a potential threat is uncovered, less privacy is provided in order to support an appropriate public health response. Another area of research is on decentralized surveillance tools and systems that can provide analysts with information about data trends and clustering, rather than the data values themselves. Such systems promise access to far more data with fewer privacy concerns.

A second approach for protecting privacy calls for audits of accesses to information and generators of linkages to deter those who might do so

---

[77]The question of what set of facts and circumstances constitutes reasonable cause belongs to the realm of policy and law rather than to technology or social science. An analogous situation exists with the privacy of medical records. While it is possible for unauthorized parties to hack their way past security systems that guard medical records, the largest and most significant disclosures of medical records to third parties occur because policy and law exist to allow such disclosures in a perfectly legal manner. See Computer Science and Telecommunications Board, National Research Council. 1997. *For the Record: Protecting Electronic Heath Information*. National Academy Press, Washington, D.C.

[78]Computer Science and Telecommunications Board, National Research Council. 2000. *Summary of a Workshop on Information Technology Research for Federal Statistics*. National Academy Press, Washington, D.C., p. 36.

inappropriately and to punish those who actually do. In this scenario, an individual uses authorized access to a computer system to obtain sensitive information or develops a sensitive linkage and then uses it for some inappropriate purpose. Research is needed to investigate the feasibility of technologies that can mitigate the damage done when "insiders" use technological means to obtain such information inappropriately and also increase the likelihood that the individual abusing his or her access will be caught. Effective organizational policies, practices, and processes to counter such abuses are also an important area of research.

A third approach—entirely based in policy—is to take steps that reduce the harm suffered by individuals whose privacy is compromised inappropriately. That is, many (though by no means not all) concerns about privacy arise from the fear that improperly disclosed information might be used to an individual's economic or legal detriment. Thus, a concern about privacy with respect to records of HIV status may be partly rooted in a fear that improper disclosure might result in the loss of health insurance or the denial of a job opportunity in the future; a concern about the privacy of one's financial records may be rooted in a fear that one could become a "mark" for criminals or the subject of unwarranted tax audits. Laws and regulations prohibiting the use of information obtained through inappropriate disclosures and providing victims with reasonable recourse should such disclosures occur could serve to ease public concerns about privacy. (Such laws and regulations are likely to be relevant also to individuals with inaccurate data associated with them, e.g., an inaccurate report of HIV status on a medical record.)

3.5 OTHER IMPORTANT TECHNOLOGY AREAS

A number of other areas of IT are important for counterterrorism purposes. While the committee believes that the research areas described in Sections 3.1 through 3.4 should have the highest priority in any comprehensive IT research program for counterterrorist purposes, the material below is intended to indicate why these areas are important and to sketch out how the research might be done.

3.5.1 Robotics

Robots are useful in the areas of military operations, hostile environments, and toxic-waste management. They can serve in high-tech arenas such as space exploration or in home environments— as personal assistants for the elderly. Autonomous vehicles in particular include robotic cars, tanks, aircraft, helicopters, land rovers, and "snakes" that crawl up and down rough surfaces. The area of robotics has made substantial advances in recent years, especially when combined with technologies such as mobile wireless, virtual reality, and intelligent or behavior-based software.

As an extension of small devices and sensors, robots serve two purposes. First, robots currently exist to assist or replace emergency workers or military personnel in dangerous situations. For example, bomb squads use robots to inspect, open, and destroy or detonate suspicious packages; tethered or wireless robots, with receivers and transmitters, could respond to commands to disarm bombs.

A second purpose for robots could be as sensing devices to detect motion and airborne chemicals. But unlike stationary sensors, robots have the ability to move. For example, a "search-and-rescue" robot could crawl into crevices that are inaccessible or dangerous for humans (e.g., gas-filled pockets), provide surveillance via a mounted camera or microphone, and communicate a situational assessment with a wireless transmitter. The use of such robots presents challenges such as their being fit into an existing organizational and information hierarchy and being used by workers who might not be familiar with technology.

Though the field of robotics has existed in some form for many years, practically useful robots are only beginning to emerge. For example, the robots used today for search and rescue are little more than remote-controlled vehicles. Much more research has to be done on making them function more autonomously, with more robust control and behavior circuits. In addition, the management of teams of robots poses important research problems with respect to multiagent learning, planning, and execution in the face of uncertainty and potential opponents.

Robots combine complex mechanical, perceptual, computer, and telecommunications systems. More work on all of these areas—and on the problems of integration—needs to be done. Robot arms and manipulators are still a major source of weakness. Robotics is a canonical example of a research area that cuts across disciplines. With research problems in pattern recognition, planning, object manipulation (especially at a fine level), machine emotions, mechanical compliance, control theory, and methods for human-robot interaction, robotics draws on expertise from and poses challenges in computer science, mechanical engineering, and electrical engineering, as well as the social sciences.

3.5.2 Sensors

Sensors are relevant in preventing certain kinds of attack (e.g., threat-warning sensors that detect smuggled nuclear materials), detecting certain kinds of attack (e.g., incident-response sensors that indicate the presence of deadly but odorless chemical agents), and mitigating the consequences of an attack (e.g., sensors that indicate whether an individual has been exposed to a biological agent). (For more discussion of particular sensor-specific challenges related to counterterrorism, see the parent report, *Making the Nation Safer*.[79] The networking of sensors will also be necessary for broad coverage to be achieved; this area also presents research challenges.[80])

Sensors can be deployed in various environments to gather data: on rooftops to detect airborne chemicals, in hallways to detect movement, and within physical infrastructure such as buildings and bridges to detect metal fatigue and points of failure. These devices could be integrated into objects or systems that are likely to last for long periods of time and must function under constraints such as limited power source, need for adequate heat dissipation, and limitations on bandwidth and memory. Moreover, because they protect a civilian rather than a military population (civilian populations are much less tolerant of false positives and much more vulnerable to false negatives), sensors for counterterrorist purposes must place a very high premium on overall accuracy (i.e., low false positive and low false negative rates).[81] Characteristics such as operation against the widest possible number of agents and wide area coverage are also important.

In addition to the challenges related to emergency sensor deployment described in Section 3.2.6, these areas pose difficult research challenges. Sensors are most effective when individual sensors are linked and coordinated in a distributed sensor network. Such a network allows information to be collected, shared, and processed via a "digital nervous system" for situational assessment and personnel monitoring. Perhaps the most important research problem for sensor networks is their self-configurability—sensor networks must be able to configure themselves (i.e., interconnect available elements into an ensemble that will perform the required functions at the desired performance level) and adapt to their environments automatically (respond to changes in the environment or in system resources). This is not to deny the problems faced in particular sensor modalities (e.g., the automatic recognition of dangerous objects in an x-ray of carry-on luggage), but it nevertheless remains a key challenge

---

[79]National Research Council. 2002. *Making the Nation Safer: The Role of Science and Technology in Countering Terrorism*. National Academies Press, Washington, D.C., pp. 320-324.

[80]CSTB, NRC, 2001, *Embedded, Everywhere*.

[81]In general, for any given test, as the criteria for a positive result are increased, the rate of false positives decreases while the rate of false negatives increases. The only way to decrease both simultaneously is to devise a better test.

to undertake sensor networking and to integrate into a single threat picture sensor readings from a large number of sensors (this is important because of the need for monitoring large areas) each of which are of low accuracy.

3.5.3 Simulation and Modeling

Models are mathematical representations of a system, entity, phenomenon, or process, while simulation is a method for implementing a model over time.[82] Modeling and simulation can play important roles throughout crisis-management activities. Specifically, emergency responders in a crisis must take actions to mitigate imminent loss of life and/or property. Thus, the ability to make predictions about how events might unfold (e.g., how a plume of chemical or biological agents might disperse in the next few hours, how a fire might spread to adjoining areas, how long a building might remain standing) is important to emergency responders. Simulations can also be useful for testing alternative operational choices.

While traditional simulation models have been applied to severe storms, earthquakes, and atmospheric dispersion of toxic substances, their primary focus has been scientific research rather than real-time crisis response. One difference between scientific simulations and crisis simulations is that the presentation of results for scientific purposes is not always compatible with the needs of emergency responders. For example, plume models of a chemical spill or release of radioactive material typically produce maps showing dispersion in parts per million as a function of time, whereas an emergency responder needs an automatic translation of the concentration of materials into easily interpretable categories such as "Safe," "Hazardous but not life threatening," or "Life threatening" so that appropriate action can be taken quickly.

Research in simulation and modeling is needed in several areas if these are to become useful tools for emergency responders:

- *Developing ad hoc models.* In many situations, a particular terrorist scenario will not have been anticipated. Thus, it would be helpful to have tools to facilitate the creation of ad hoc situation-specific models rapidly (i.e., on a time scale of hours or days). This would include both the data-

---

[82]The discussion in this section is adapted largely from CSTB, NRC, 1996, *Computing and Communications in the Extreme,* and CSTB, NRC, 1999, *Information Technology Research for Crisis Management.*

gathering needed for input to the model and the construction of the model itself. Such tools would likely incorporate significant amounts of subject-matter expertise. The performance of such models could not be expected to match that of models developed under leisurely conditions, but they might provide rough guidance that could be very useful to emergency responders. (As with other issues relating to the use of technology, how people actually use such tools matters a great deal. Thus, understanding the actual utility of hastily constructed models as a function of the practice and experience of the users involved is also an essential component of this research.)

- *Increasing the timeliness of modeling results.* Because of the time pressures facing emergency responders, timely results are critical. One element is fast processing capabilities. Another element, easy to overlook, is the timeliness of data that can be inserted into a model. To be useful for emergency responders, data must be entered on a near-real-time basis (e.g., a model could be linked to weather sensor data), a task that becomes impossible if large efforts at data preprocessing or formatting are necessary before a simulation can be run.

- *Obtaining adequate computational power.* Because many simulations require large amounts of computing power, techniques are needed that enable the rapid requisitioning of computers engaged in scientific research and other activities to serve simulation needs in a crisis. Such techniques will require both new administrative arrangements and further advances in the flexibility, affordability, and ease of use of these resources.

- *Rapid rescaling of models.* Different emergency responders may need predictions at different time and distance scales. For example, the mayor of a city might require information on whether to order an evacuation in response to a successful attack on a nuclear plant and thus would need information on what might happen at the plant on a time scale of 24 hours. On the other hand, a firefighter at the plant may need to know what might happen in the next hour. The ability to run simulations at varying scales of resolution is important for both the mayor and the firefighter, and the needs of each drive the trade-offs between the accuracy of a prediction and the speed of the response. Thus, research is needed to develop the capability for rescaling models rapidly in response to requests from crisis managers.

- *Making initial damage estimates.* Rapid assessment of the extent and distribution of damage in the wake of any large incident is difficult because acquiring and synthesizing damage reports take considerable time. However, initial damage estimates are essential for directing response efforts. For example, the destruction of a dam could release a tidal wave of water that would destroy some houses and spare others. A model that provided initial damage estimates would have to account for building stock (structure type, age, and so on), critical facilities, and lifelines, as well as geological information and demographics, in order to predict the number of casualties and the need for shelter and hospitals. Note also that such models would have direct relevance to the insurance industry and might help guide the development of criteria and pricing for terrorism insurance policies.

- *Improving model interoperability.* As a rule, models are developed and used in isolation. Integrating models into information systems for emergency responders would enable more accurate and timely predictions to be made. For example, emergency responders at the Murrah Building in Oklahoma City in 1995 used computer-aided design (CAD) software to map the areas to be searched and to correlate estimated locations of victims (based on where their offices had been located before the blast) with the actual scene. More useful would have been the coupling of the CAD data into a structural model that could perform finite-element analysis to predict the loads on various parts of the damaged building, thus indicating where shoring was necessary to prop up damaged structures and reduce the danger to survivors and rescuers from further collapses.

3.6 PEOPLE AND ORGANIZATIONS

The craft of espionage distinguishes between the initial penetration of an organization and an ongoing exploitation that continues to yield useful information or access. Technology (e.g., a worm, rogue code, a bug, a planted microchip) is often essential to the ongoing exploitation, but it is almost always social methods (e.g., bribing a low-level worker) that allow the initial penetration which facilitates the introduction of that technology. For example, a bug planted in a room will reveal conversations there for a long time, but it is a failure of people and organizations that allows an enemy agent to plant the bug in the first place. In the immediate aftermath of World War II, Soviet spies in the United States used a one-time pad encryption system, an encryption approach that is provably unbreakable. Nevertheless, U.S. cryptanalysts broke the Soviet code—because the Soviets began to reuse some of the key pads.[83]

---

[83]Key pads are simply lists of random numbers, so there was no particular reason why the Soviets were forced to reuse them. Their reuse was a human error that violated the essential premise underlying the security of one-time pad encryption systems. (See NSA

This lesson generalizes to any study that involves the weaknesses and vulnerabilities of any system. That is, technology is always used in some social and organizational context, and human error and human culpability are central in understanding how the system might fail. Thus, design and deployment are system issues, and since human, social, and organizational behavior are part of the system, they must be part of the research and design efforts. The technology cannot be examined in isolation of the ways in which it is deployed.

Technology is aimed at helping people, organizations, and society to accomplish their goals. Technology is essential to modern everyday life, and technology is essential to thwarting would-be terrorists and criminals. At the same time, if deployed poorly, the technology can actually make problems worse. If it is poorly designed, it will lead to numerous errors, usually blamed on the unwitting users of the system, but almost invariably traceable to poor design from a human or organizational point of view. Whatever the reason, human error can be extremely costly in time, money, and lives. Research findings from the behavioral and social sciences can aid in the design of solutions that are effective while being minimally obtrusive. Good design can dramatically reduce the incidence of error—as the experience in commercial aviation shows.

3.6.1 Principles of Human-Centered Design

Both technology and the people around it make errors. It is simply not possible to have zero false positives and zero false negatives simultaneously, especially in a world filled with uncertainty, ambiguity, and noise. Most importantly, deliberate adversaries seek to cause and exploit false alarms. As a result, the design of a system involving both technology and people must assume some percentage of false positives and false negatives. The design of a system must provide multiple levels of defense, accommodate continual feedback, and allow systematic attempts to improve. In most cases, most errors should be taken as useful feedback and used to improve system performance rather than as signs of failure that require punishment and blame.

The best systems will look for incipient failures—problems that are detected before they do damage—as important measures of operation.

Venona papers online at <http://www.nsa.gov/docs/venona/monographs/monograph-2.html>.) More generally, the use of one-time pad systems is cumbersome and difficult and does not scale easily to large numbers of users, so they are disfavored for many applications. These disadvantages are all human factors and illustrate well the trade-off between security and ease of use.

The time to take corrective action is when incipient errors reach an undesirable state but before they turn into serious issues. It is only natural to ignore these early warnings because, of course, no damage has yet been done. But it was exactly this ignoring of early-warning trouble signs that led directly to the disaster with the O-rings on the Space Shuttle *Challenger*.[84]

In general, systems must be designed from a holistic, systems-oriented point of view rather than by focusing on the technology. Indeed, understanding the human and organizational interactions is important well before technical requirements are established. Then, as the system is being designed, the experts on technology and human behavior jointly devise possible systems and develop rapid prototypes that allow quick and efficient testing of the ideas. Attention to the human requirements early in the design stage coupled with iterative design techniques saves time in the end by minimizing the end testing requirements and reducing the likelihood of major revisions late in the development process.

Consider the phenomenon of "human error." Experience demonstrates that often what is blamed on "human error" is in fact design error. A focus on the technology in isolation of the manner in which it is deployed invariably leads to problems. It is common in the development of automated systems to automate whatever can be done, leaving the rest to the human operators. This often means that the human is "out of the loop," with little responsibility, until matters reach the point at which the automatic systems can no longer cope. Then the human is suddenly and unexpectedly faced with an emergency, and because at this point everything is under human control, any failure is labeled "human error."

By contrast, when systems are designed with a full understanding of the powers—and weaknesses—of human operators, the incidence of human error is greatly diminished. People must be given meaningful tasks. They must always be engaged (i.e., "in the loop"), and their talents should be employed for high-level guidance, not for entering detailed sequences that require high accuracy. Machines are good at accurate, repetitive actions. People are not.

Human-centered design is a well-explored field with several societies, professional journals, and much academic research as well as applications, especially in aviation safety and computer systems.[85] The following subsections address some basic principles of human-centered design.

---

[84]See R.P. Feynman. 1986. "Personal Observations on Reliability of Shuttle," in App. F, *Report of the Presidential Commission on the Space Shuttle Challenger Accident*. Available online at <http://science.ksc.nasa.gov/shuttle/missions/51-l/docs/rogers-commission/Appendix-F.txt>. Accessed November 18, 2002.

[85]See, for example, Jef Raskin, 2000, *The Human Interface*, Addison-Wesley, Boston.

Defend in Depth and Try to Avoid Common-Mode Failures

Many of today's systems have a single point of defense. If problems arise there, if there is some error, there is no second line to fall back on.[86] A notable exception is aviation safety, which builds multiple systems so that failure at one point does not break the system.

A key difficulty in such a design philosophy is the existence of common-mode failures. The accident literature is filled with examples of common-mode failure, in which redundant safety systems all failed at the same time. For example, in the September 11 attack on the World Trade Center, the medical response was hindered by the fact that the city's Office of Emergency Management (OEM), responsible for coordinating all aspects of a disaster response, was housed in Building 7 of the World Trade Center, one of the buildings that collapsed several hours after the airplane strikes.

This is not to say that common-mode failures are usually ignored in design. But they can often be difficult to detect and eliminate. The "redundant" communications lines cited in Section 2.2.2 are a potential common-mode failure, and yet as the discussion of Section 3.2.8 emphasizes, redundant links that are in fact not redundant are often not easy to identify. A more complex example of a hidden common-mode failure occurred in a commercial airliner that lost oil pressure in all three engines simultaneously. The two engines on the wing were quite different from the engine in the tail, and they had been serviced by different technicians. However, it turned out that each engine required an O-ring seal, and due to a complex chain of events, the O-ring was left off the part that was serviced on all three engines. Thus, in this case, the O-ring was the common-mode failure, though on the surface the simultaneous failure did not appear to be a common-mode failure.[87]

Account for the Difference Between Work as Practiced and Work as Prescribed

In the study of work practices, it is commonplace to observe the distinction between work as practiced and work as prescribed. When people describe their work, they can provide clear and coherent statements about

---

[86]P.A. Hancock and S.G. Hart. 2002. "Defeating Terrorism: What Can Human Factors/Ergonomics Offer?" *Ergonomics in Design*, Vol. 10 (1): 6-16.

[87]National Transportation Safety Board (NTSB). 1984. *Aircraft Accident Report—Eastern Air Lines, Inc., Lockheed L-1011, N334EA, Miami International Airport, Miami, Florida, May 5, 1983*. Report No. NTSB/AAR-84/04. National Transportation Safety Board, Washington, D.C.

the procedures they follow. But observation of these same people doing actual work shows that they are inconsistent with the descriptions they provided. Why? Because "that was a special case." In fact, it turns out that special cases are the norm, and the descriptions are of the prototypical case that seldom actually happens. Additionally, these descriptions of procedure are for nonemergency circumstances. During emergencies, established procedures will prove even more inadequate than they are under putatively normal circumstances.

A common approach to improving safety and security is to tighten procedures and to require redundant checking. But tighter procedures to improve security address work "as prescribed." In practice, the technology and procedures that are added to make operations safer and more secure quite often get in the way of getting the work done. Security technology and procedures can introduce so many problems into getting the job done that people learn to circumvent them. Because people are inherently helpful and well motivated to do their work, they develop workarounds to bypass security, not because they are not well trained or motivated, but precisely because they *are* well trained and motivated. In many cases, they could not accomplish their tasks without violating procedures. This is especially true in crises, where normal routines break down and workarounds are essential.

Consider, for example, advice that is often given about passwords. People are advised to use passwords that are long and obscure and to change them frequently so that if one is compromised, it does not remain compromised for long. The result is that people write down their passwords on yellow Post-it notes and paste them on their terminals, where they are easy for unauthorized users to see.

Biometrics provides another example. Widely accepted as a stronger authentication mechanism than passwords, even biometrics cannot be the entire solution. Biometric mechanisms must sometimes be bypassed for entirely legitimate reasons. "I burned my hand and it's all bandaged up, so I can't use the fingerprint machine. Can you let me in just this once?" "I just had a cataract operation and I have to wear an eye patch, so how do I do an iris scan?" Any opening for legitimately bypassing normal procedures opens the door for illegitimate bypasses. Even if technology cannot be fooled (and it almost always can be), the people behind the technology can be fooled.

Sometimes problems occur because the pressures on individuals differ from the stated goals of the organization. When people are asked to follow arduous security requirements while at the same time maintaining efficient and productive work schedules, there can be conflicts. Workers must choose which is of greatest importance. Quite often it is the work schedule that is given priority, and although this is not unreasonable, it occurs at the cost of security. The proper goal is to design systems and procedures in which these are not in conflict.

In short, there is a terrible trade-off: the very things that make security more secure are often those that make our lives more difficult, or in some cases, impossible. When human and organizational factors are not taken into account in system design and development, the measures intended to increase security may reduce it because dedicated, concerned workers will thwart such measures in order to get their jobs done. Realistic security is cognizant of human and organizational behavior.

Plan for People Who Want to Be Helpful

Crooks, thieves, criminals, and terrorists are experts at exploiting the willingness of people to be helpful—a process usually known as "social engineering." These adversaries use people to help them understand how to use the onerous technology, and they use people by taking advantage of situations that cause breakdowns in normal procedures. In short, they help human error to occur.

For example, badges are often required for entry into a secure facility. However, entry can usually be obtained in the following manner: Walk up to the door carrying an armload of computers, parts, and dangling cords. Ask someone to hold the door open, and thank them. Carry the junk over to an empty cubicle, look for the password and log-in name that will be on a Post-it note somewhere, and log in. If you cannot log in, ask someone for help. As one guide for hackers puts it, "Just shout, 'Does anyone remember the password for this terminal?' . . . you would be surprised how many people will tell you."[88]

A firefighter who needed to know the security code to get in a secure building through the back door might call the management office of the building and say:

> Hey, this is Lt. John Hennessy from Firehouse 17. This is a Code 73 emergency. There are people screaming behind the doors, the building is going to collapse—what's the security code? Hurry, I'm losing my signal. Hello? Hello? Better hurry. [Sounds of screaming and other loud noises in background.]

A terrorist who needed to know the security code to enter a secure building through the back door would call and say:

---

[88] "The Complete Social Engineering FAQ." Available online at <http://morehouse.org/hin/blckcrwl/hack/soceng.txt>.

Hey, this is Lt. John Hennessy from Firehouse 17. This is a Code 73 emergency. There are people screaming behind the doors, the building is going to collapse—what's the security code? Hurry, I'm losing my signal. Hello? Hello? Better hurry. [Sounds of screaming and other loud noises in background.]

Times of stress and emergency, when security is perhaps of most importance, are exactly the times when the strains are the greatest and the need for normally nonauthorized people (such as firefighters, police, rescue and health personnel) to gain access is acute. And this is when it is easiest for a terrorist to get in, using the same mechanisms.

Countering social engineering by an adversary is an important counterterrorist technique. But whatever the counter, the solution must not be based on extinguishing the tendencies of people to be helpful. The reason is that helpful people often play a key role in getting any work done at all—and thus the research challenge is to develop effective techniques for countering social engineering that do not require wholesale attacks on tendencies to be helpful.

Understand Bystander Apathy

As more people are involved in checking a task, it is possible for safety to decrease. This is called the "bystander apathy" problem, named after studies of a New York City crime in which numerous people witnessed an incident but no one helped and no one reported it. Laboratory studies showed that the greater the number of people watching, the lower the likelihood that anyone would help—the major reason being that each individual assumes that if an incident is serious, someone else out of all those watching will be doing something, so the fact that nobody is doing anything means that it must not be a real issue. After all, in New York, anything might be happening: it might be a movie shoot. Similarly, if I am asked to check the meter readings of a technician, and I know that the immediate supervisor has already checked them and that someone else will check my report, I don't take the check all that seriously: after all, how could a mistake get through with so many people involved? I don't have to worry. But what if everyone feels that way?

The commercial aviation community has done an excellent job of fighting this tendency with its program of crew resource management (CRM). In CRM, the pilot not flying is required to be an active critic of the actions taken by the pilot who is flying. And the pilot flying is supposed to thank the other for the criticism, even when it is incorrect. Obviously, getting this process in place was difficult, for it involved major changes in the culture, especially when one pilot was junior, the other very senior. But the result has been increased safety in the cockpit.

Account for Cognitive and Perceptual Biases

The research literature in cognitive and social psychology clearly demonstrates that most people are particularly bad at understanding low-probability events. This might be the "one-in-a-million" problem. An airplane pilot might well decide that the situation of three different oil pressure indicators all reading zero oil pressure is likely fallacious, because it is a "one-in-a-million" chance that all three engines would fail at the same time. However, since there are roughly 10 million commercial flights a year in the United States, one-in-a-million means that 10 flights a year will suffer from this problem.

On the other hand, salient events are overestimated in frequency. Thus, aviation is considered more dangerous than automobile driving by many people, despite the data that show just the opposite. What about terrorist acts? These are truly unlikely, deadly though each may be, but if airline passengers overreact, they are apt to attack and possibly seriously injure an innocent passenger who meets some of their preconceptions of a terrorist. Indeed, each successful encounter between passengers and potential harrowers increases the likelihood of a future false encounter.

The "boy who cried wolf" is a third perceptual bias—potential threats are often ignored because of a history of false alarms. An effective criminal or terrorist approach is to trigger an alarm system repeatedly so that the security personnel, in frustration over the repeated false alarms, either disable or ignore it—which is when the terrorist sneaks in.

Probe and Test the System Independently

The terms "red team" and "tiger team" refer to efforts undertaken by an organization to test its security from an operational perspective using teams that simulate what a determined attacker might do. Tiger teams develop expertise relevant to their intended targets, conduct reconnaissance to search for security weaknesses, and then launch attacks that exploit those weaknesses. Under most circumstances, the attack is not intended to be disruptive but rather to indicate what damage could have been done. Properly conducted tiger-team testing has the following characteristics:

- It is conducted on an unscheduled basis without the knowledge of the installation being probed, so that a realistic security posture can be tested.

- It does not function under unrealistic constraints about what it can or cannot do, so its attack can realistically simulate what a real attacker might do.
- It reports its results to individuals who are not directly responsible for an installation's security posture so that negative results cannot be suppressed.
- It probes and tests the fundamental assumptions on which security planning is based and seeks to violate them in order to create unexpected attacks.

Why are tiger teams a "people and organization" issue? The essential reason is that an attacker has the opportunity to attack any vulnerable point in a system's defenses, whether that point of vulnerability is the result of an unknown software bug, a misconfigured access control list, a password taped to a terminal, lax guards at the entrance to a building, or a system operator trying to be helpful.

Over the years, tiger teams have been an essential aspect of any security program, and tiger-team tests are essential for several reasons:

- Recognized vulnerabilities are not always corrected, and known fixes are frequently found not to have been applied as a result of poor configuration management.
- Security features are often turned off in an effort to improve operational efficiency. Such actions may improve operational efficiency, but at the potentially high cost of compromising security, sometimes with the primary damage occurring in some distant part of the system.
- Some security measures rely on procedural measures and thus depend on proper training and ongoing vigilance on the part of commanders and system managers.
- Security flaws that are not apparent to the defender undergoing an inspection may be uncovered by a committed attacker (as they would be uncovered in an actual attack).[89]

In order to maximize the impact of these tests, reports should be disseminated widely. The release of such information may cause embarrassment of certain parties or identify paths through which adversaries may attack, but especially in the case of vulnerabilities uncovered for which fixes are available, the benefits of releasing such information—allowing others to learn from it and motivating fixes to be installed—generally outweigh these costs. Furthermore, actions can be taken to mini-

---

[89]CSTB, NRC, 1999, *Realizing the Potential of C4I*, p. 147.

mize the possibility that adversaries might be able to obtain or use such information. For example, passing the information to the tested installation using nonelectronic means would eliminate the possibility that an adversary monitoring electronic channels could obtain it. Delaying the public release of such information for a period of time could allow the vulnerable party to fix the problems identified.

Finally, tiger-team testing launched without the knowledge of the attacked systems also allows estimates to be made of the frequency of attacks. Specifically, the fraction of tiger-team attacks that are detected is a reasonable estimate of the fraction of adversary attacks that are made. Thus, the frequency of adversary attacks can be estimated from the number of adversary attacks that are detected.

3.6.2 Organizational Practices in IT-Enabled Companies and Agencies

An organization's practices play an important role in countering external threats. The discussion below is not meant in any way to be exhaustive but rather to motivate examination of yet another nontechnological dimension of security.

Outsourcing of Product Development and Support

For entirely understandable reasons, many companies outsource IT work to parties whose interests may not be fully aligned with their own. Companies outsource for many reasons, ranging from the availability of skilled human resources that are not indigenous to them to the often-lower cost of doing so (especially when the parties doing the outsourced work have access to cheaper sources of labor).

The practice of outsourcing has security implications. On the one hand, outsourced work represents a potential vulnerability to the company that uses such work, unless that company has the expertise to audit and inspect the work for security flaws. By assumption, a company that outsources work has less control over how the work is done, and the possibility of deliberately introduced security vulnerabilities in outsourced work must be taken seriously. On the other hand, one reason for outsourcing work is that those doing the work may have greater expertise than the company hiring them—and if security is a special expertise of the former, its capabilities for maintaining security may be *greater* than that of the latter.

Outsourcing is not in and of itself a practice that leads to insecure IT systems and networks. Nevertheless, prudence dictates that a company understand the potential risks and benefits of outsourcing from a security perspective. If it does outsource work, a company should undertake careful and informed inspection of the work on system components that provide critical functionality.

Personnel Screening

Behavioral and psychological profiles of typical outside "hackers" have been available for a long time, providing insight into their motivations and techniques. However, similar information about persons likely to present an insider threat is not available today. One challenge to assembling such information is the fact that insider adversaries can be characterized in many different ways. For example, the behavior of the insider will likely vary depending on a wide variety of factors, including whether the person is unwitting, incompetent, coerced, vengeful, and so on. Such factors imply that simply relying on externally observable traits and behaviors in order to identify potential insiders may not prove useful, and so integration with background information on individual employees may be necessary to identify potential risks from insiders. Note also that all screening techniques run the risk of incorrectly labeling problematic behavior acceptable or of determining benign behavioral patterns to be indicative of inappropriate behavior or intent.

Managing Personnel in a Security-Oriented Environment

Managing employees in a security-oriented environment is complex. The practices that characterize the handling of classified information often impede the sharing of information among people. Long-term compliance with security procedures is often difficult to obtain, as employees develop ways to circumvent these procedures in order to achieve greater efficiency or effectiveness. Personnel matters that are routine in nonsecure environments become difficult. For example, from a security perspective, termination of the access privileges of employees found to be improperly hired or retained must happen without warning them of such termination. On the other hand, due process may prevent rapid action from being taken. The temptations are strong to relax the requirements of due process for security, but not observing due process often has detrimental effects on organizational morale and esprit de corps,[90] not to mention the possibility that almost any pretext will suffice for some individual supervisors to eliminate workers they do not personally like.

---

[90]Illustrations of the conflict between applying due process and managing the requirements of security can be found in the cases of Wen Ho Lee and Felix Bloch. In investigating the alleged passing of nuclear weapons design information to the People's Republic of

Many issues arise when an employee is merely under suspicion for wrongdoing or malicious intent—before action is taken to rescind access privileges, the company may run a significant risk of suffering significant damage. This suggests that under these circumstances, the work of the employee must be monitored and controlled, but at the same time the requirements of due process must be observed.

3.6.3 Dealing with Organizational Resistance to Interagency Cooperation

An effective response to a serious terrorist incident will inevitably require the multiple emergency-response agencies to cooperate. Section 3.2.1 describes technical barriers to effective cooperation, but technological limitations by no means explain why agencies might fail to cooperate effectively.

Specifically, it is necessary to note that the character and traditions of agencies have a profound impact on their ability and willingness to cooperate. Different agencies exhibit many differences in internal cultures (e.g., in philosophies of staff reward and punishment, in traditions among disciplines in research and implementation, in ethical criteria of staff in terms of private versus public interest, in performance criteria between public and profit-making enterprises, and in degree of participatory decision making). Turf battles and jurisdictional warfare between agencies with overlapping responsibilities are also common, with each agency having its own beliefs about what is best for the citizenry.

For the public record, the rhetoric of every emergency-response agency acknowledges the need for cooperation with other agencies. But the reality in practice is often quite different from the rhetoric, and in practice almost every disaster (whether natural or attack-related) reveals shortcomings in the extent and nature of interagency cooperation.

China, Lee was charged on multiple counts of mishandling material containing restricted data with the intent to injure the United States and with the intent to secure an advantage to a foreign nation. After being charged, Lee was held in custody under conditions described by the cognizant federal judge as draconian, because it was believed that his pretrial release would pose a grave threat to the nation's security. The case ended with the dismissal of all but one charge of mishandling classified information and with the judge's apology for the conduct of the government in the prosecution of this case. Felix Bloch was a Foreign Service officer in the State Department, investigated in 1989 by the FBI for spying for the Soviet Union. Bloch was eventually fired and stripped of his pension in 1990 on grounds that he lied to FBI investigators, but he was never charged. It is alleged that Robert Hanssen gave information to the Soviets revealing that Bloch was under suspicion, which might account for the fact that sufficient grounds for charging Bloch were never found.

For example, the emergency response to the September 11, 2001, attacks on the World Trade Center revealed a number of cultural barriers to cooperation in the New York City Police and Fire Departments:[91]

- Police helicopters, with the ability to provide firefighters close-up information on the progress of the fire in the upper parts of the buildings as well as some aerial rescue capability of those gathered on the roofs, were never used for those purposes. An on-site Fire Department chief tried to request police helicopters for such roles but was unable to reach the police dispatcher for the helicopters either by phone or radio. Furthermore, the Fire Department had established its command post in the building lobbies, while the police had established their command post three blocks away, and the police did not report to the Fire Department commanders on-site. Said one senior Fire Department official, "They [the police] report to nobody and they go and do whatever they want."
- The Police and Fire Departments have a formal agreement (in place since 1993) to share police helicopters during high-rise fires and to practice together. However, neither agency has any records of joint drills, although some less formal "familiarization flights" may have been conducted for the Fire Department a year or so before September 11.
- While most states and the federal government have forged agreements among emergency-response agencies that specify in advance who will be in overall charge of a crisis response, New York City has no such agreement, which left its Police and Fire Departments with no guidance about how to proceed with overall command arrangements on September 11.
- Police fault firefighters and firefighters fault police for unwillingness to cooperate. Some police believe that sharing command with the Fire Department is difficult because firefighters lack paramilitary discipline characteristic of the police force, while some firefighters thought that the police felt they could and should do everything.
- In the aftermath, senior Fire Department and Police Department officials disagreed over the extent to which the departments were able to coordinate. A senior Fire Department official said that "there is no question there were communications problems [between the Fire Department and the Police Department] at this catastrophic incident," while a senior Police Department official said, "I was not made aware that day that we were having any difficulty coordinating."

---

[91]See Jim Dwyer, Kevin Flynn, and Ford Fessenden. 2002. "9/11 Exposed Deadly Flaws in Rescue Plan." *New York Times*, July 7. Available online at <http://www.nytimes.com/2002/07/07/nyregion/07EMER.html?pagewanted=1>.

Cultural barriers separating the CIA and the FBI have also been revealed in the postmortems that have been conducted since September 11. In particular, the essential mission of the CIA is one of intelligence collection and analysis, while the essential mission of the FBI has been directed toward law enforcement. As a broad generalization, these missions differ in that intelligence is more focused on anticipating and predicting bad events, while law enforcement is more focused on prosecutions and holding perpetrators of dangerous events accountable to a criminal justice system. To illustrate, intelligence analysts place a high value on protecting sources and methods for gathering intelligence so that they will be able to continue obtaining information from those channels, while law-enforcement officials place a high value on the ability to use information in open court to gain convictions. (The fact that the FBI and CIA also operate under very different legal regimes governing their domestic activities is also quite relevant, but beyond the scope of this report. Suffice it to say that these different legal regimes impose explicit behavioral constraints and serve to shape the environment in which the cultural attitudes within each agency develop.)

Desires of agencies to preserve their autonomy also contribute to an active (if subterranean) resistance to interoperability. Personnel of one agency without the capability of communicating with another agency are not easily directed by that other agency. Furthermore, an agency may have a fear (often justified) that communications overheard by another agency will lead to criticism and second-guessing about actions that it took in the heat of an emergency.

There are no easy answers for bridging cultural gulfs between agencies that do not interact very much during normal operations. Different agencies with different histories, different missions, and different day-to-day work would be expected to develop different policies, procedures, and philosophies about what is or is not appropriate under a given set of circumstances.

For the most part and under most circumstances, an agency's culture serves it well. But in crisis, interagency differences impede interagency cooperation, and they cannot be overcome by fiat at the scene of the crisis. For example, a policy directive requiring that agencies adopt and use common communications protocols does not necessarily require emergency responders from different agencies to actually interact with one another while an emergency response is occurring.

3.6.4 Principles into Practice

Putting these principles into practice requires that the human requirements be considered equally with technical and security requirements.

The most secure and reliable systems will be those developed with behavioral scientists from the user interface community who use an iterative design-test-design implementation strategy.

The major point is to recognize that security and reliability are systems problems: the needs and standard working practices of the people involved are as important as the technical requirements. The very virtues of people are often turned against them when intruders seek to broach security: the willingness to help others in distress is perhaps the weakest link of all in any defensive system, but it would be preferable to design security systems that detract minimally from this valuable human attribute. Many failures are due to security requirements that are unreasonable from the point of view of human cognition (e.g., asking for frequent memorization of long, complex passwords) or that severely impact the ability to get the required work done because they conflict with the organizational structures and requirements. It is therefore essential that the needs of the individuals, the workgroups, and the organization all be taken into account. Conscientious workers will do whatever is necessary to get the work done, often at the cost of compromising security. But through proper design, it should be possible to design systems that are both more efficient and more secure.

To achieve effective interagency cooperation in crisis, many things must happen prior to the occurrence of crisis, taking into account the realities of organizational resistance to interoperability. Such cooperation is likely to require:

• *Strong, sustained leadership.* When a strong leader places a high priority on interagency cooperation, is willing to expend resources and political capital in support of such cooperation, and can sustain that expenditure over a time long enough so that the agencies in question cannot "outwait" his or her efforts, organizational change that moves in the direction of that priority is more likely.

• *Activities that promote interagency understanding and cooperation.* One example of such activities is the temporary detailing of personnel from one agency to another (e.g., firefighters and police officers or FBI and CIA analysts on temporary duty at each other's agencies). Prior exposure to one another's operational culture generally helps to reduce frictions that are caused by lack of familiarity during a crisis. Of course, to be genuinely helpful, this practice must be carried out on a sufficiently large scale that the personnel so exposed are likely to be those participating in a response to a crisis. Another useful activity is joint exercises that simulate crisis response. As a rule, exercises that involve most or all of the agencies likely to be responding in a disaster—and that use the IT infrastructure that they are expected to use—are an essential preparation for effective interagency cooperation.[92] Exercises help to identify and solve some social, organizational, and technical problems, and they help to reveal the rivalries and infighting between agencies whose resolution is important to real progress in this area. To the extent that the agencies and personnel involved in an exercise are the same as those involved in the response to a real incident, exercises can help the response to be less ad hoc and more systematic. Other, less formal activities can also be conducted to improve interagency understanding. For example, personnel from one agency can be detailed to work in another agency in emergency-response situations. As a part of pre-service and in-service training, personnel from one agency can be posted to other agencies for short periods to develop contacts and to understand the operating procedures of those other agencies.

- *Budgets that support interagency cooperation.* For many agencies, battles of the budget are as important as their day-to-day operational responsibilities. This is not inappropriate, as adequate budget resources are a prerequisite for an agency's success. Thus, it is simply unrealistic to demand cooperation from agencies without providing budget resources that are dedicated to that end. Note that budget resources support operations (e.g., personnel and training matters) and the procurement of systems, and both are relevant to interagency cooperation.

3.6.5 Implications for Research

The discussion in the previous sections has two purposes. One is to describe the operational milieu into which technology is deployed—a warning that human beings are an essential part of any operational system and that system design must incorporate sophisticated knowledge of human and organizational issues as well as technical knowledge. A second purpose is to develop a rationale for research into human and organizational issues relating to technology in a counterterrorism context.

Research in this area will be more applied than basic. The social sciences (used broadly to include psychology, anthropology, sociology, organizational behavior, human factors, and so on) have developed a significant base of knowledge that is relevant to the deployment of IT-based systems. But in practice, social scientists with the relevant domain expertise often lack applied skills and the requisite technical IT knowl-

---

[92]Despite the creation of New York City's Office of Emergency Management in 1996 and expenditures of nearly $25 million to coordinate emergency response, the city had not conducted an emergency exercise between 1996 and September 11, 2001, at the World Trade Center—which had been bombed in 1993—that included the Fire Department, the police, and the Port Authority's emergency staff.

edge. Similarly, information technologists often lack the appropriate domain knowledge and often use a system development process that makes it difficult to incorporate human and organizational considerations.

This point suggests that research is relevant in at least four different areas.

1. *The formulation of system development methods that are more amenable to the incorporation of domain knowledge and social science expertise.* The "spiral development" methodology for software development is an example of how user inputs and concerns can be used to drive the development process, but the method is hard to generalize to incorporate knowledge about the organizational context of use.

2. *The translation of social science research findings into guidelines and methods that are readily applied by the technical community.* The results of this research effort might very well be software toolboxes as well as a "Handbook of Applied Social Science" or a "Cognitive Engineering Handbook" containing useful principles for system development and design derived from the social science research base.

3. *The development of reliable security measures that do not interfere with what legitimate workers must do.* These methods must minimize loads on human memory and attention and task interference while providing the appropriate levels of security in the face of adversaries who use sophisticated technologies as well as social engineering techniques to penetrate the security.

4. *Understanding of the IT issues related to the disparate organizational cultures of agencies that will be fused under the Department of Homeland Security.* This is a complex task, with difficult technical issues interspersed with complex procedural, permissions, and organizational issues requiring a mix of technical and social skills to manage. Operationally, the question is how to allow for the sharing of communication and data among different organizations that have different needs to know, differing requirements, and different cultural and organizational structures, in a way that enhances the desired goals while maintaining the required security.

4

What Can Be Done Now?

Developing a significantly less vulnerable information infrastructure is an important long-term goal for the United States. This long-term goal must focus on the creation of new technologies and paradigms for enhancing security and reducing the impact of security breaches. In the short term, the committee believes that the vulnerabilities in the communications and computing infrastructure of the first-responder network should receive focused attention. Efforts should concentrate on hardening first responders' communications capability as well as those portions of their computing systems devoted to coordination and control of an emergency response. The committee believes that existing technology can be used to achieve many of the needed improvements in both the telecommunications and computing infrastructures of first responders. Unfortunately, the expertise to achieve a more secure system often does not reside within the host organizations—this may be the case, for example, in local and state government. These facts lead to two short-term recommendations.

Short-Term Recommendation 1: The nation should develop a program that focuses on the communications and computing needs of emergency responders. Such a program would have two essential functions:

• *Ensuring that authoritative current-knowledge expertise and support regarding information technology are available to emergency-response agencies prior to and during emergencies, including terrorist attacks.* One implementation option is to situate the mechanism administratively in existing government or private organizations—for example, the National Institute of Standards and Technology, the Department of Homeland Security, the Department of Defense, the Computer Emergency Response Team of the Software Engineering Institute at Carnegie Mellon University. A second option is to create a national body to coordinate the private sector and local, state, and federal authorities.[1] In the short term, a practical option for providing emergency operational support would be to exploit IT expertise in the private sector, much as the armed services draw on the private sector (National Guard and reserve forces) to augment active-duty forces during emergencies. Such a strategy, however, must provide adequate security vetting for private-sector individuals serving in this emergency role and must also be a complement to a more enduring mechanism for providing ongoing IT expertise and assistance to emergency-response agencies.

• *Upgrading the capabilities of the command, control, communications, and intelligence (C3I) systems of emergency-response agencies through the use of existing technologies and perhaps minor enhancements to them.* One key element of such upgrading should be a transition from legacy analog C3I systems to digital systems. Of course, in the short term, this transition can only be started, but it is clear that it will be necessary over the long term to achieve effective communications capabilities. In addition, maintaining effective communications capability in the wake of a terrorist attack is a high priority, and some possible options for implementing this recommendation include a separate emergency-response communications network that is deployed in the immediate aftermath of a disaster and the use of the public network to support virtual private networks, with priority given to traffic from emergency responders. (Table 4.1 describes some illustrative advantages and disadvantages of each approach.) Given the fact that emergency-response agencies are largely state and local, there is no federal agency that has the responsibility and authority over state and local responding agencies needed to carry out this recommendation. Thus, it is likely that a program of this nature would have to rely on incentives (probably financial) to persuade state and local responders to participate and to acquire new interoperable C3I systems.

---

[1]CSTB has a pending full-scale project on information and network security R&D that will address federal funding and structure in much greater detail than is possible in this report. See the Web site <http://www.cstb.org> for more information on this subject.

TABLE 4.1 A Comparison of Separate Emergency Networks with Reliance on Surviving Residual Capacity

Emergency Network	Illustrative Advantage	Illustrative Disadvantage
Separate network deployed after an emergency	Provides high-confidence assurance of known bandwidth availability.	Would not be the system regularly used by personnel; without continuous updates and training, they may not be able to use it properly in emergency settings. Deployment of network may take too long.
Residual public-network capacity plus priority for emergency responders	Assures immediate availability of some bandwidth because some part of the public network is likely to survive any disaster.	Not possible to assure the availability of adequate bandwidth for emergency responders because availability depends on the amount of surviving public network.

Short-Term Recommendation 2: The nation should promote the use of best practices in information and network security in all relevant public agencies and private organizations. Nearly all organizations, whether in government or the private sector, could do much better with respect to information and network security than they do today, simply by exploiting what is already known about that subject today, as discussed at length in *Cybersecurity Today and Tomorrow: Pay Now or Pay Later*.[2] Users of IT, vendors in the IT sector, and makers of public policy can all take security-enhancing actions.

Users of IT in individual organizations are where the "rubber meets the road"—they are the people who must actually make the needed changes work. Only changes in operational practice and deployed technology in individual organizations can have an impact on security, and

---

[2]Computer Science and Telecommunications Board (CSTB), National Research Council (NRC). 2002. *Cybersecurity Today and Tomorrow: Pay Now or Pay Later*. National Academy Press, Washington, D.C. (hereafter cited as CSTB, NRC, 2002, *Cybersecurity Today and Tomorrow*). The discussion in that volume is based on extensive elaboration and analysis contained in various CSTB reports. including *Computers at Risk* (1991), *Trust in Cyberspace* (1999), and *Realizing the Potential for C4I* (1999), among others.

the parties responsible for taking action range from chief technical (or even executive) officers to system administrators. Individual organizations can and should:

- Establish and provide adequate resources to an internal entity with responsibility for providing direct defensive operational support to system administrators throughout the organization . . . . To serve as the focal point for operational change, such an entity must have the authority—as well as a person in charge—to force corrective action.
- Ensure that adequate information-security tools are available, that everyone is properly trained in their use, and that enough time is available to use them properly. Then hold all personnel accountable for their information system security practices . . . .
- Conduct frequent, unannounced red-team [tiger-team] penetration testing of deployed systems and report the results to responsible management . . . .
- Promptly fix problems and vulnerabilities that are known or that are discovered to exist . . . .
- Mandate the organization-wide use of currently available network/configuration management tools, and demand better tools from vendors . . . .
- Mandate the use of strong authentication mechanisms to protect sensitive or critical information and systems . . . .
- Use defense in depth. In particular, design systems under the assumption that they will be connected to a compromised network or a network that is under attack, and practice operating these systems under this assumption.
- Define a fallback plan for more secure operation when under attack and rehearse it regularly. Complement that plan with a disaster-recovery program.[3]

Vendors of IT systems and services have key roles to play in improving the security functionality of their products. Such vendors should:

- Drastically improve the user interface to security, which is [virtually] incomprehensible in nearly all of today's systems . . . . Users and administrators must be able to easily see the current security state of their systems; this means that the state must be expressible in simple terms.
- Develop tools to monitor systems automatically for consistency with defined secure configurations, and enforce these configurations. . . . Extensive automation is essential to reduce the amount of human labor

---

[3]CSTB, NRC, 2002, *Cybersecurity Today and Tomorrow*, p. 13.

that goes into security. The tools must promptly and automatically respond to changes that result from new attacks.
- Provide well-engineered schemes for user authentication based on hardware tokens . . . . These systems should be both more secure and more convenient for users than are current password systems.
- Develop a few simple and clear blueprints for secure operation that users can follow, since most organizations lack the expertise to do this properly on their own. For example, systems should be shipped with security features turned on, so that a conscious effort is needed to disable them, and with default identifications and passwords turned off, so that a conscious effort is needed to select them . . . .
- . . . [c]onduct more rigorous testing of software and systems for security flaws, doing so before releasing products rather than use customers as implicit beta testers to . . . [uncover] security flaws . . . .[4] Changing this mind-set is one necessary element of an improved . . . posture [for information and network security].[5]

In addition, vendors should provide individual consumers with easy-to-use, default-on security tools and features to secure home computers and networks. Because home computers can play a significant role in attacks against cyber infrastructure, actions securing this diffuse infrastructure could help to reduce the potential threat it poses.

Makers of public policy have an important role in securing critical government IT systems and networks. The Office of Management and Budget (OMB) has sought to promote government information and network security in the past, but despite its actions, the state of information and network security in government agencies remains highly inadequate. In this regard, the administration and Congress can position the federal government as a leader in technology use and practice by requiring agencies to adhere to the practices recommended above and to report on their progress in implementing those measures.[6] Such a step would also help to grow the market for security technology, training, and other services.

---

[4]"Note that security-specific testing of software goes beyond looking at flaws that emerge in the course of ordinary usage in an Internet-connected production environment. For example, security-specific testing may involve very sophisticated attacks that are not widely known in the broader Internet hacker community."

[5]CSTB, NRC, 2002, *Cybersecurity Today and Tomorrow*, pp. 13-14.

[6]This concept has been implicit in a series of laws, beginning with the Computer Security Act of 1987, and administrative guidance (e.g., from OMB and more recently from the Federal Chief Information Officers Council). Although it has been an elusive goal, movements toward e-government have provided practical, legal, and administrative impetus. For more discussion, see Computer Science and Telecommunications Board, National Research Council. 2002. *Information Technology Research, Innovation, and E-Government*. National Academy Press, Washington, D.C.

> **BOX 4.1 A Comparison of Fire Safety with Information and Network Security**
>
> Today's fire codes seek to provide a certain level of safety against fire in buildings, and there is broad acceptance of the idea that compliance with fire codes results in buildings that are safer against fire than those that are not compliant. Fire codes are also developed and enforced by government regulation. A reasonable question is, Why can't the same kind of regulation be used to improve information and network security?
>
> Fire safety and information and network security are very different in certain key dimensions:
>
> - *Intentionality.* Most fires are accidental, and hence the fire code is not primarily concerned with the deliberate bypassing of fire safety measures. Arson presents a very different problem (fortunately rare compared to the accidents that account for the majority of fires), and if arson were the primary problem in fire safety, fire codes would look very different indeed—and would likely be much less effective at making buildings safer than they are in today's environment. However, in the area of information and network security, most system penetrations are deliberate, and so information and network security is much more like protecting against arson than protecting against accidental fires.
> - *Monoculture versus diversity.* A building code seeks to standardize the construction of buildings in ways that improve fire resistance. But standardization regarding safety measures in buildings is useful only when the threats to buildings (in this case, the threat of fires in each of the buildings in question) are independent and uncorrelated. That is, the ways in which fires can start are highly varied, and so measures that are ineffective against one type of fire may well be useful against another type. In the case of information and network security for a largely homogeneous environment, the threat is highly correlated—an attacker who develops techniques for penetrating the security measures of one system knows how to penetrate the security measures of many.
> - *The rate of change in the underlying technologies.* Buildings have existed for many years, and fire has been known to be a threat for a long time. Buildings take

As for the private sector, there is today no clear locus of responsibility within government to undertake the "promotion" of security across the private sector, because neither information and network security in the private sector nor IT products and services are subject today to direct government regulation.[7] This will not necessarily always be true, but for

---

[7] In this context, "direct regulation" is taken to mean government-issued mandates about what the private sector must do with respect to cybersecurity.

a long time to design and construct, and the techniques for designing and constructing them are relatively stable. By contrast, information technology changes rapidly. Thus, the computer and network systems being protected change quickly, their vulnerabilities change quickly, and the threat changes quickly.

- *Visibility of damage.* As a rule, fires create visible damage. But the damage to a computer system or a network may be entirely invisible; indeed, a system that fails to operate normally is only one possible result of an attack on it. A successful attack may lay the foundation for later attacks (e.g., by installing Trojan horse programs that can be subsequently activated), or it may be set to cause damage well after the initial penetration, or enable the clandestine and unauthorized transmission of sensitive information stored on the attacked system (e.g., password files).
- *The underlying science.* The science underlying fire safety is much better understood and developed than that underlying information and network security. For example, it is understood how to build a fire-resistant structure from first principles. One might specify the use of steel beams that lose structural integrity at a certain temperature. Finite-element analysis based on a sound underlying mathematics enables reliable predictions to be made about structural loading. But no such science underlies information and network security and the development of secure systems and networks.
- *The availability of metrics.* In fire codes, it is meaningful to specify that a building must resist burning for a certain period of time. But there is no comparable metric to specify how long a computer system or network must be able to resist an intruder. More generally, there is no quantitative basis for understanding how much security is made available by the addition of any particular feature in computer or network design.

These important differences should not be taken to mean that nothing is known about information and network security—and as discussed in the main text, there are common sense measures that can be taken that do improve such security. But direct regulation is always more difficult to impose when the benefits are uncertain and/or difficult to articulate, and for this reason, those who wish to impose direct regulation to improve information and network security face many difficulties that warrant thought and deliberation.

a number of reasons (described in Box 4.1), the realities of information and network security make it less amenable to government regulation than other fields such as fire or automobile or flight safety.

In addition, the committee notes that the IT sector is one over which the federal government has little leverage. IT sales to the government are a small fraction of the IT sector's overall revenue, and because IT purchasers are generally unwilling to acquire security features at the expense of performance or ease of use, IT vendors have little incentive to include security features at the behest of government alone. Indeed, it is likely that attempts at such regulation will be fought vigorously, or may fail, because of the likely inability of a regulatory process to keep pace with rapid changes in technology.

Thus, appropriate market mechanisms could be more successful than direct regulation in improving the security of the nation's IT infrastructure, even though the market has largely failed to provide sufficient incentives for the private sector to take adequate action with respect to information and network security. The challenge for public policy is to ensure that those appropriate market mechanisms develop. How to deal constructively with prevailing market dynamics has been an enduring challenge for the government, which has attempted a variety of programs aimed at stimulating supply and demand but which has yet to arrive at an approach with significant impact.

Nevertheless, the committee believes that public policy can have an important influence on the environment in which nongovernment organizations live up to their responsibilities for security. One critical dimension of influencing security-related change is the federal government's nonregulatory role, particularly in its undertaking of research and development of the types described above.[8] Such R&D might improve security and interoperability, for example, and reduce the costs of implementing such features—thereby making it less painful for vendors to adopt them.

Other policy responses to the failure of existing incentives to cause the market to respond adequately to the security challenge are more controversial. If the market were succeeding, there would be a significant private sector demand for more security in IT products, and various IT vendors would emphasize their security functionality as a competitive advantage and product differentiator, much as additional functionality and faster performance are featured today. But this is not the case. Possible options to alter market dynamics in this area include:

- Increasing the exposure of software and system vendors and system operators to liability for system breaches;[9]

---

[8] Another potentially important aspect of the government's nonregulatory role, a topic outside the scope of this report, is the leadership role that government itself could play with respect to information and network security. For more discussion, see CSTB, NRC, 2002, *Cybersecurity Today and Tomorrow*.

[9] CSTB, NRC, 2002, *Cybersecurity Today and Tomorrow*.

[10] CSTB, NRC, 2002, *Cybersecurity Today and Tomorrow*.

- Mandatory reporting of security breaches that could threaten critical societal functions;[10]
- Changing accounting procedures to require sanitized summaries of information-security problems and vulnerabilities to be made public in shareholder reports; and
- Encouraging insurance companies to grant preferential rates to companies whose IT operations are regarded as meeting certain security standards of practice.

Note, however, that there are disadvantages as well as advantages to any of these specific options, and a net assessment of their ultimate desirability remains to be undertaken.

5

Rationalizing the Future Research Agenda

As noted in Chapter 3, the committee believes that the IT research areas of highest priority for counterterrorism are in three major areas: information and network security,[1] information technologies for emergency response, and technologies for information fusion. Within each of these areas, a reasonably broad agenda is appropriate, as none of them can be characterized by the presence of a single stumbling block or impediment whose removal would allow everything else to fall into place.

Attention to human and organizational issues in a counterterrorism context is also critical. Insight, knowledge, and tools that result from such attention are likely to be much more relevant to systems integration than to technology efforts devoted to proofs-of-principle or other technology development issues. However, that fact does not mean that there is no role for research, especially since system development methodologies that incorporate such tools are scarce or nonexistent. Thus, the engagement of social scientists (e.g., psychologists, anthropologists, sociologists, organizational behavior analysts) will be important in any research program in IT for counterterrorist purposes.

Based on the discussion in Chapter 3, Box 5.1 summarizes some of the

---

[1]Further discussion of a broader research agenda on information and network security can be found in CSTB's *Computers at Risk* (1991) and *Trust in Cyberspace* (1999). Though these reports were issued several years ago, their comments on a relevant research agenda remain pertinent today, reflecting the reality that the information-security field has not advanced much in the intervening years.

topics within these areas that the committee believes would be fruitful to research. It is useful to note that progress in these areas would have commercial applications as well in many cases. The fruits of information and network security research would benefit all users of information technology, though their particular relevance to providers of critical infrastructure is obvious. Emergency responders will be the primary beneficiaries of research that focuses on their particular needs. Progress in information fusion has relevance across the spectrum of counterterrorism efforts, from prevention to detection to response, and indeed to information mining for other public and private purposes. (A point of particular interest is the fact that information-fusion efforts for countering bioterrorism have significant applicability to public health, especially with respect to the early identification of "natural" disease outbreaks.) Advances in developing tools to incorporate knowledge about human and organizational factors in systems integration would be relevant to the deployment of most large IT-based systems.

The fact that research in these areas may have commercial relevance raises for some questions about the necessity of government involvement. As noted in Chapter 4, the commercial market has largely failed in promoting information and network security. In other cases, the research program required (e.g., research addressing the needs of emergency responders) is of an applied nature—and focused on counterterror applications. As for information fusion, it is highly likely that its applications will have commercial applications once new technologies are developed, but whether those new technologies would develop in the absence of government-supported research and become broadly available is another question entirely.

Most of these technology research areas are not new. Efforts have long been under way in information and network security and information fusion, though additional research is needed because the resulting technologies are not sufficiently robust or effective, they degrade performance or functionality too severely, or they are too hard to use or too expensive to deploy. Moreover, given the failure of the market to adequately address security challenges, adequate government support for R&D in information systems and network security is especially important. Information technologies for emergency response have not received a great deal of attention, though efforts in other contexts (e.g., military operations) are intimately related to progress in this area.[2]

---

[2]Military communications and civilian emergency-response communications have similarities and differences. Military forces and civilian agencies share the need to deploy emergency capacity rapidly, to interoperate, and to operate in a chaotic environment. But while military communications must typically work in a jamming environment or one in

> **BOX 5.1 Illustrative Topic Areas for Long-Term Research**
>
> *Authentication, Detection, Identification*
>
> - Develop fast and scalable methods for high-confidence authentication.
> - Explore approaches that could self-monitor traffic and users to detect either anomalous users or unusual traffic patterns.
> - Develop intruder-detection methods that scale to function efficiently in large systems.
>
> *Containment*
>
> - Develop the tools and design methodologies for systems and networks that support graceful degradation in response to an attack.
> - Develop mechanisms to contain attackers and limit damage rather than completely shutting down the system once an intrusion is detected.
> - Explore how to fuse a simple, basic control system used during "crisis mode" with a sophisticated control system used during normal operations.
>
> *Recovery*
>
> - Develop schemes for backing up large systems, in real time and under "hostile" conditions, that can capture the most up-to-date, but correct, snapshot of the system state.
> - Create new decontamination approaches for discarding as little good data as possible and for removing active and potential infections on a system that cannot be shut down for decontamination.
>
> *Cross-cutting Issues in Information and Network Security*
>
> - Develop tools that support security-oriented systems development.
> - Find new ways to test bug fixes reliably.
> - Develop better system-administration tools for specifying security policies and checking against prespecified system configurations.
> - Create new tools to detect added and unauthorized functionality.
> - Develop authentication mechanisms that provide greater security and are easier to use.
> - Create and employ metrics to determine the improvement to system security resulting from the installation of a security measure.
> - Monitor and track emerging types of attack and explore potential consequences of such attacks.
> - Understand why previous attempts to build secure systems have failed and recommend how new efforts should be structured to be more successful.
>
> *C3I Systems for Emergency Response*
>
> - Understand how to transition gracefully and with minimal disruption from a unit-specific communication system to a systemwide structure.
> - Define new communication protocols and develop generic technology to facilitate interconnection and interoperation of diverse information sources.
> - Develop approaches for communication systems to handle surge capacity and function in a saturated state.
> - Develop methods to provide more capacity for emergency communication and coordination.
> - Create self-adaptive networks that can reconfigure themselves as a function of damage and changes in demand and that can degrade gracefully.

- Understand the special security needs of rapidly deployed wireless networks.
- Develop decision-support tools to assist the crisis manager in making decisions based on incomplete information.
- Explore mechanisms to provide information tailored to specific individuals or locations through location-based services.
- Establish more effective means of communicating the status of affected people to those outside the disaster area.
- Develop robust sensors and underlying architectural concepts to track and locate survivors as well as to identify and track the spread of contaminants.
- Create digital floor plans and maps of other physical infrastructure, and use wearable computers and "map ants" to generate maps that can be updated.
- Develop tools to map network topology, especially of converged networks that handle voice and data traffic.
- Begin to characterize the functionality of regional networks for emergency responders.

Information Fusion for Counterterrorism

- Develop more effective machine-learning algorithms for data mining, including learning for different data types (text, image, audio, video).
- Develop methods for systems to learn when data are scarce.
- Create better mixed-initiative methods that allow the user to visualize the data and direct the data analysis.
- Explore new methods to normalize and combine data from multiple sources.
- Create methods to extract structured information from text.
- Build approaches to handle multiple languages.
- Improve algorithms for image interpretation, speech recognition, and interpretation of other sensors (including perception based on mixed media).
- Extend, and test extensively in more demanding applications, the principle-based methods for reasoning under uncertainty.
- Develop techniques for machine-aided query formulation.
- Develop visualization techniques that are well-adapted for unstructured data.

Privacy and Confidentiality

- Understand the impact on confidentiality of different kinds of data disclosure.
- Develop data-mining algorithms that can be used without requiring full disclosure of individual data records.

Human and Organizational Factors

- Create system development methods that more easily accommodate inputs relevant to human and organizational factors.
- Develop software toolboxes and handbooks that codify and encapsulate principles derived from the social sciences that are relevant to system development and design.
- Develop reliable security measures that do not interfere with legitimate workers.
- Understand the IT issues related to the disparate organizational cultures of agencies that will be fused under the Department of Homeland Security.

NOTE: A future CSTB report on cybersecurity research will explicate research areas in greater detail.

As for the funding of the research program described in this report, computer crime losses are estimated at $10 billion per year (and growing).[3] Although statistics on the amount lost to cybercrime are of dubious reliability, there is no doubt that aggregate losses are considerable. The committee believes that because this research program has considerable overlap with that needed to fight cybercrime, progress in this research program has the potential to reduce cybercrime as well. Without rigorous argument, the committee believes that the potential reduction in cybercrime would likely offset a considerable portion (if not all) of the research program described in this report (though of course the primary beneficiaries will be society at large rather than any individual company that today may suffer loss). Nevertheless, the committee has not had access to information that would allow it to determine an appropriate level of funding for the research program described in this report.

The time scale on which the fruits of efforts in these research areas will become available ranges from short to long. That is, each of these areas has technologies that can be beneficially deployed on a relatively short time scale (e.g., in a few years). Each area also has other prospects for research and deployment on a much longer time scale (e.g., a decade or more) that will require the development of entirely new technologies and capabilities.

The committee is silent on the specific government agency or agencies that would be best suited to support the program described above,[4] though it notes that the recently created Department of Homeland Security may expand the options available for government action. Rather, the more important policy issue is how to organize a federal infrastructure to support this research. In particular, the committee believes that this infrastructure should have the following attributes. It would:

• Engage and support multidisciplinary, problem-oriented research that is useful both to civilian and military users. (Note that this approach contrasts strongly with the disciplinary orientation that characterizes most academic departments and universities.)
• Develop a research program driven by a deep understanding and

---

which there is a need for a low probability of intercept, these conditions do not obtain for civilian emergency-response communications. Also, military forces often must communicate in territory without a pre-existing friendly infrastructure, while civilian agencies can potentially take advantage of such an infrastructure.

[3]"Cyber Crime." *BusinessWeek Online*, February 21, 2000. Available online at <http://www.businessweek.com/2000/00_08/b3669001.htm>.

[4]See CSTB, NRC, 2002, *Cybersecurity Today and Tomorrow*, pp. 13-14.

assessment of IT vulnerabilities. This will likely require access to classified information, even though most of the research should be unclassified.

- Support a substantial effort in research areas with a long time horizon for payoff. Historically, such investigations have been housed most often in academia, which can conduct research with fewer pressures for immediate delivery on a bottom line. (This is not to say that private industry has no role. Indeed, because the involvement of industry is critical for deployment, and is likely to be essential for developing prototypes and mounting field demonstrations, it is highly appropriate to support both academia and industry perhaps even jointly in efforts oriented toward development.)

- Provide support extending for time scales that are long enough to make meaningful progress on hard problems (perhaps 5-year project durations) and in sufficient amounts that reasonably realistic operating environments for the technology could be constructed (perhaps $2 million to $5 million per year per site for system-oriented research programs).

- Invest some small fraction of its budget on thinking "outside the box" in consideration (and possible creation) of alternative futures (Box 5.2).

- Be more tolerant of research directions that do not appear to promise immediate applicability. Research programs, especially in IT, are often—even generally—more "messy" than research managers would like. The desire to terminate unproductive lines of inquiry is understandable, and sometimes entirely necessary, in a constrained budget environment. On the other hand, it is frequently very hard to distinguish between (A) a line of inquiry that will never be productive and (B) one that may take some time and determined effort to be productive. While an intellectually robust research program must be expected to go down some blind alleys occasionally, the current political environment typically punishes such blind alleys as being of Type A, with little apparent regard for the possibility that they might be Type B.

- Be overseen by a board or other entity with sufficient stature to attract top talent to work in the field, to provide useful feedback, and to be an effective sounding board for that talent.

- Pay attention to the human resources needed to sustain the counterterrorism IT research program. This need is especially apparent in the fields of information and network security and emergency communications. Only a very small fraction of the nation's graduating doctoral students in IT specialize in either of these fields, only a very few professors conduct research in these areas, only a very few universities support research programs in these fields, and, in the judgment of the committee,

> **BOX 5.2 Planning for the Future**
>
> Planning for the future is a critical dimension of any research agenda, though the resources devoted to it need not be large. System architectures and technologies such as switched optical networks, mobile code, and open-source or multinational code development will have different vulnerabilities from the technologies that characterize most of the existing infrastructure and hence require different defense strategies. Similarly, device types such as digital appliances, wireless headphones, and network-capable cell phones may pose new challenges. Even today, it is hard to interconnect systems with different security models or security semantics; unless this problem is successfully managed, it will become increasingly difficult in the future.
>
> Furthermore, the characteristics of deployed technology that protect the nation against catastrophic IT-only attacks today (e.g., redundancy, system heterogeneity, and a reliance on networks other than the Internet for critical business functions) may not continue to protect it in the future. For example, trends toward deregulation are pushing the nation's critical infrastructure providers to reduce excess capacity, even though this is what provides much of the redundancy so important to reduced vulnerability. In the limit, the market dominance of a smaller number of products leads to system monocultures that, like their ecological and agricultural counterparts, are highly vulnerable to certain types of attack.
>
> For these reasons, researchers and practitioners must be vigilant to changes in network technology, usage and reliance on IT, and decreasing diversity. In addition, research focused on the future is likely to have a slant that differs from the orientation of the other research efforts described in this chapter. While the latter efforts might be characterized as building on existing bodies of knowledge (and are in that sense incremental), future-oriented research would have a more radical orientation: it would, for example, try to develop alternative paradigms for secure and reliable operation that would not necessarily be straightforward evolutions from the Internet and information technology of today. One such pursuit might be the design of appropriate network infrastructure for deployment in 2020 that would be much more secure than the Internet of today. Another might be an IT infrastructure whose security relied on engineered system diversity—in which deployed systems were sufficiently similar to be interoperable, yet sufficiently diverse to essentially be resistant to large-scale attacks.

only a very small fraction of the universities that do support such programs can be regarded as first-rate universities.

One additional attribute of this R&D infrastructure would be desirable, though the committee has few good ideas on how to achieve it. The success of the nation's R&D enterprise in IT (as well as in other fields) rests in no small part on the ability of researchers to learn from each other in a relatively free and open intellectual environment. Constraining the openness of that environment (e.g., by requiring that research be classified or by forbidding certain research from being undertaken) would have obvious negative consequences for researchers and the creation of new knowledge. On the other hand, keeping counterterrorist missions in mind, the free and open dissemination of information has potential costs as well, because terrorists may obtain information that they can use against us. Historically, these competing interests have been "balanced"— with more of one in exchange for less of the other. But the committee believes (or at least hopes) that there are other ways of reconciling the undeniable tension, and calls for some thought to be given to a solution to this dilemma that does not demand such a trade-off. If such a solution can be found, it should be a design characteristic of the R&D infrastructure.

A comment on the counterterrorist research program is that successfully addressing the privacy and confidentiality issues that arise in counterterrorism efforts will be critical for the deployment of many information technologies. This area is so important that research in the area itself is necessary and should be a fundamental component of the work in virtually all of the other areas described in this report.

Finally, it is the belief of the committee that an R&D infrastructure with the characteristics presented above has the best chance of delivering successfully on the complex research problems described in this report. The committee is not arguing for unlimited latitude to undertake research that is driven primarily by intellectual curiosity, but rather for a program focused on the specific national needs described in this report that can look beyond immediate deliverables. More detailed research agendas should be forthcoming from the agencies responsible for implementing the broad research program described in this report.

Appendix

Biographies of Committee and Staff Members

COMMITTEE ON THE ROLE OF INFORMATION TECHNOLOGY
IN RESPONDING TO TERRORISM

John L. Hennessy, Chair, is president of Stanford University, where he joined the faculty in 1977, was the chair of the Department of Computer Science in 1994, and became dean of the School of Engineering in 1996. He is an expert in computer architecture and is recognized for innovation in software techniques as the codeveloper of reduced instruction set computing (RISC). In 2001 Dr. Hennessy received the Eckert-Mauchly Award from the Association for Computing Machinery and Institute of Electrical and Electronics Engineers Computer Society and honorary doctoral degrees from the State University of New York at Stony Brook and Villanova University, his two alma maters. He is currently chairman of the board at Atheros, and was a board member at Alentec Corporation and an advisory board member at Microsoft Corporation and Tensilica. Over the past decade, he has served on numerous committees at the National Academies, most recently as the chair of the Computer Science and Engineering Committee and the Committee on Membership during 1999-2000. He has served on the Computer Science and Telecommunications Board committees that produced *Global Trends in Computer Technology and Their Impact on Export Control, Academic Careers for Experimental Computer Scientists and Engineers,* and *Evolving the High Performance Computing and Communications Initiative to Support the Nation's Information Infrastructure.* He has also provided his computer expertise and leadership skills on committees and commissions of the National

Science Foundation and the Defense Advanced Research Projects Agency (DARPA). He is an alumnus of CSTB and a member of Tau Beta Pi, Eta Kappa Nu, Pi Mu Epsilon, and the National Academy of Engineering.

David A. Patterson, *Vice Chair*, is the E.H. and M.E. Pardee Chair of Computer Science at the University of California at Berkeley. He has taught computer architecture since joining the faculty in 1977 and has been chair of the Computer Science Division of the Electrical Engineering and Computer Science Department at Berkeley. He is well known for leading the design and implementation of RISC I, the first Very Large-Scale Integration (VLSI) Reduced Instruction Set Computer, which became the foundation for the architecture currently used by Fujitsu, Sun Microsystems, and Xerox. He was also a leader of the Redundant Arrays of Inexpensive Disks (RAID) project, which led to high-performance storage systems from many companies, and the Network of Workstation (NOW) project, which led to cluster technology used by Internet companies such as Inktomi. He is a fellow of the Institute of Electrical and Electronics Engineers and the Association for Computing Machinery. He served as chair of the Computing Research Association. His current research interests are in building novel microprocessors using Intelligent Dynamic Random Access Memory (IRAM) for use in portable multimedia devices and using Recovery Oriented Computing to design available, maintainable, and evolvable servers for Internet services. He has consulted for many companies, including Digital Equipment Corporation, Hewlett Packard, Intel, and Sun Microsystems, and he is the coauthor of five books. Dr. Patterson served on the CSTB committees that produced *Computing the Future: A Broader Agenda for Computer Science and Engineering* and *Making IT Better: Expanding Information Technology Research to Meet Society's Needs*. He is a member of the National Academy of Engineering and a current member of CSTB.

Steven M. Bellovin, fellow at AT&T Research, is a renowned authority on security—in particular, Internet security. Dr. Bellovin received a B.A. degree from Columbia University and an M.S. and Ph.D. in computer science from the University of North Carolina at Chapel Hill. While a graduate student, he helped create Netnews; for this, he and the other collaborators were awarded the 1995 USENIX Lifetime Achievement Award. At AT&T Laboratories, Dr. Bellovin does research in networks and security, and why the two do not get along. He has embraced a number of public interest causes and weighed in (e.g., through his writings) on initiatives (e.g., in the areas of cryptography and law enforcement) that appear to compromise privacy. He is currently focusing on cryptographic protocols and network management. Dr. Bellovin is the coauthor of the recent book *Firewalls and Internet Security: Repelling the Wily Hacker*, and he is a member of the Internet Architecture Board. He has recently been elected to the National Academy of Engineering. He served on the CSTB committee that produced *Trust in Cyberspace* and is a member of the committee to study authentication technologies and their implications for privacy. That committee is addressing a range of security (and privacy) issues, including those relating to data collection (e.g., biometrics) and analysis (e.g., tracking systems), as well as a range of systems issues.

W. Earl Boebert is an expert on information security, with experience in national security and intelligence as well as commercial applications and needs. He is a senior scientist at Sandia National Laboratories. He has 30 years experience in communications and computer security, is the holder or co-holder of 13 patents, and has participated in National Research Council studies on security matters. Prior to joining Sandia, he was the technical founder and chief scientist of Secure Computing Corporation, where he developed the Sidewinder security server, a system that currently protects several thousand sites. Before that he worked 22 years at Honeywell, rising to the position of senior research fellow. At Honeywell he worked on secure systems, cryptographic devices, flight software, and a variety of real-time simulation and control systems, and won Honeywell's highest award for technical achievement for his part in developing a very large scale radar landmass simulator. He also developed and presented a course on systems engineering and project management that was eventually given to more than 3,000 students in 13 countries. He served on the CSTB committees that produced *Computers at Risk: Safe Computing in the Information Age* and *Trust in Cyberspace*, and participated in Project Initiation Fund workshops on "Cyber-Attack" and "Insider Threat."

David Borth is an expert on wireless communications, with insight into national security as well as commercial needs. He is corporate vice president and director of the Communication Systems and Technologies Laboratory of Motorola Incorporated, a part of the company's research arm, Motorola Laboratory. Dr. Borth joined Motorola in 1980 as a member of the Systems Research Laboratory in corporate research and development in Schaumburg, Illinois. As a member of that organization, he has conducted research on digital modulation techniques, adaptive digital signal processing methods applied to communication systems, and personal communication systems including both cellular and Personal Communications Service systems. He has contributed to Motorola's implementations of the GSM, TDMA (IS-54/IS-136), and CDMA (IS-95) digital cellular systems. In his current role, he manages a multinational (United States, Australia, France, Japan) organization focusing on all aspects of communication systems, ranging from theoretical systems studies to system and subsystem analysis and implementation to integrated circuit designs. Dr. Borth received his B.S., M.S., and Ph.D. degrees in electrical engineering from the University of Illinois at Urbana-Champaign. Previously, he was a member of the technical staff of the systems division of Watkins-Johnson Company and an assistant professor in the School of Electrical Engineering, Georgia Institute of Technology. Dr. Borth is a member of Motorola's Science Advisory Board Associates and has been elected a Dan Noble Fellow, Motorola's highest honorary technical award. He has been issued 31 patents and has authored or coauthored chapters of 5 books in addition to 25 publications. He received the Distinguished Alumnus Award from the University of Illinois Electrical and Computer Engineering Alumni Association and was elected a fellow of the Institute of Electrical and Electronics Engineers for his contributions to the design and development of wireless telecommunication systems. He is a registered professional engineer in the State of Illinois and a current member of CSTB.

William J. Brinkman has been vice president of Physical Sciences Research at Bell Labs of Lucent Technologies since 1993. He is an expert in the area of condensed matter physics as it pertains to telecommunications and information-processing technologies. He received his Ph.D. in physics from the University of Missouri in 1965. Dr. Brinkman joined Bell Laboratories in 1966 after spending one year as a National Science Foundation postdoctoral fellow at Oxford University. He moved to Sandia National Laboratories in 1984, but returned to Bell Laboratories in 1987 to become executive director of the Physics Research Division. His responsibilities include the direction of research in physical sciences, optoelectronic and electronic devices, fiber optics, and related areas. He has worked on theories of condensed matter, and his early work also involved the theory of spin fluctuations in metals and other highly correlated Fermi liquids. Subsequent theoretical work on liquid crystals and incommensurate systems are additional important contributions that he has made to the theoretical understanding of condensed matter. As manager of an industrial research organization with a budget of $200 million. Dr. Brinkman is strongly interested in improving technology conversion and improving the connection between research and products. He was the recipient of the 1994 George E. Pake Prize. Over the past 20 years, he has served on numerous committees of the National Academies, currently the National Academy of Sciences 2002 Nominating Committee and most recently the Committee on Science, Engineering, and Public Policy. He is a member of the National Academy of Sciences.

John M. Cioffi is an expert on communications technologies, with an emphasis on wireline contexts. He received his BS in electrical engineering from the University of Illinois in 1978 and his Ph.D. in electrical engineering from Stanford University in 1984. He worked for Bell Laboratories from 1978 to 1984, IBM Research from 1984 to 1986, and has been an electrical engineering professor at Stanford University since 1986. Dr. Cioffi founded Amati Communications Corporation in 1991 (purchased by Texas Instruments in 1997) and was an officer/director from 1991 to 1997. Currently he is on the boards or advisory boards of BigBand Networks, Coppercom, GoDigital, Ikanos, Ionospan, Ishoni, IteX, Marvell, Kestrel, Charter Ventures, and Portview Ventures. Dr. Cioffi's specific interests are in the area of high-performance digital transmission. He has received various awards: member, National Academy of Engineering (2001), IEEE Kobayashi Medal (2001), IEEE Millennium Medal (2000), IEEE Fellow (1996), IEE JJ Tomson Medal (2000), 1999 University of Illinois Outstanding Alumnus, 1991 IEEE *Communications Magazine* best paper, 1995 American National Standards Institute T1 Outstanding Achievement Award, National Science Foundation Presidential Investigator (1987-1992). Dr. Cioffi has published more than 200 papers and holds over 40 patents, most of which are widely licensed, including basic patents on DMT, VDSL, and vectored transmission. He served on the CSTB committee that produced *Broadband: Bringing Home the Bits* and is a current member of CSTB.

W. Bruce Croft is chair of the computer science department, as well as distinguished university professor, at the University of Massachusetts, Amherst, which he joined in 1979. In 1992, he became the director of the National Science Foundation State/Industry/University Collaborative Research Center for Intelligent Information Retrieval (CIIR), which combines basic research with technology transfer to a variety of government and industry partners. Dr. Croft received his B.Sc. (Honors) degree in 1973 and an M.Sc. in computer science in 1974 from Monash University in Melbourne, Australia. He earned his Ph.D. in computer science from the University of Cambridge, England, in 1979. His research interests are in several areas of information retrieval, including retrieval models, Web search engines, cross-lingual retrieval, distributed search, question answering, text summarization, and text data mining. He has published more than 120 articles on these subjects. Dr. Croft has consulted for many companies and government agencies. He co-founded a search engine startup in 1996, and his research is being used in a number of operational systems. He was chair of the ACM Special Interest Group on Information Retrieval from 1987 to 1991. He is currently editor-in-chief of the ACM's *Transactions on Information Systems* and an associate editor for *Information Processing and Management*. He has served on numerous program committees and has been involved in the organization of many workshops and conferences. He was elected a fellow of ACM in 1997 and received the Research Award from the American Society for Information Science and Technology in 2000. He is a member of CSTB's Digital Government committee, which has provided insight into government application contexts. He has a history of research interactions with the intelligence community, and his emphases relate to data collection and analysis. He is a current member of CSTB.

William P. Crowell is a former deputy director and chief operating officer of the National Security Agency. Prior to the NSA, he was vice president of Atlantic Aerospace Electronics Corporation. In 1998 he joined Cylink and is currently president and CEO. The company helped pioneer the use of computer security systems within major financial and government institutions. He holds a bachelor's degree in political science from Louisiana State University.

Jeffrey M. Jaffe has broad knowledge of systems, with emphases on networked/distributed systems and the associated security challenges. He is vice president of research and advanced technologies for Lucent Technologies Bell Laboratories. The Advanced Technologies Group works with Lucent's business units in the commercial development and deployment of new technologies, with an emphasis in networks planning, software and systems engineering. Prior to joining Lucent, Dr. Jaffe held a variety of research and management positions with International Business Machines (IBM). He joined IBM's Thomas J. Watson Research Center in 1979, conducting research on networking protocols. He led research teams in developing networking and security software and user interfaces. He was later promoted to a number of executive positions, including vice president of systems and software. In this role, he coordinated the efforts of global research teams in supporting IBM's current product lines and developing new software and hardware systems. Dr. Jaffe next served as corporate vice president of technology and helped to convert research into commercial products. He played key roles in assessing new technologies and policy enactment. In his most recent position with IBM, Dr. Jaffe managed all facets of IBM's network software and security product business. Dr. Jaffe is a fellow of the IEEE and the ACM. The U.S. government has consulted with him on numerous policy initiatives with a focus on the Internet. In 1997, President Clinton appointed Dr. Jaffe to the advisory committee for the President's Commission for Critical Infrastructure Protection. Dr. Jaffe has chaired the Chief Technology Officer Group of the Computer Systems Policy Project (CSPP), which consists of a dozen of the top computer and telecommunications companies. Dr. Jaffe earned a B.S. degree in mathematics, as well as M.S. and Ph.D. degrees in computer science, from the Massachusetts Institute of Technology. He is a current member of CSTB.

Butler W. Lampson is known for his expertise in systems and systems architecture. At present, he is a distinguished engineer at Microsoft Corporation, where he works on problems of broad concern, such as security and information management. Before joining Microsoft, Dr. Lampson was a senior corporate consulting engineer at Digital Equipment Corporation and a senior research fellow at the Xerox Palo Alto Research Center. He has worked on computer architecture, local area networks, raster printers, page description languages, operating systems, remote procedure call, programming languages and their semantics, programming in the large, fault-tolerant computing, computer security, and WYSIWYG editors. He was one of the designers of the SDS 940 timesharing system, the Alto personal distributed computing system, the Xerox 9700 laser printer, two-phase commit protocols, the Autonet LAN, and several programming languages. He received a Ph.D. in electrical engineering and computer science from the University of California at Berkeley and honorary Sc.D. degrees from the Eidgenoessische Technische Hochschule, Zurich, and the University of Bologna. He holds a number of patents on networks, security, raster printing, and transaction processing. Dr. Lampson is a member of the National Academy of Engineering. He received the Association for Computing Machinery's Software Systems Award in 1984 for his work on the Alto, and the Turing Award in 1992. He served on the CSTB committees that produced *Computers at Risk: Safe Computing in the Information Age*, *Evolving the High Performance Computing and Communications Initiative to Support the Nation's Information Infrastructure*, and *Realizing the Potential of C4I: Fundamental Challenges*. He is a current member of CSTB.

Edward D. Lazowska has broad knowledge of software and distributed and high-performance systems. He holds the Bill and Melinda Gates Chair in Computer Science in the Department of Computer Science and Engineering at the University of Washington. Dr. Lazowska received his A.B. from Brown University in 1972 and his Ph.D. from the University of Toronto in 1977. He has been at the University of Washington since that time. His research concerns the design and analysis of distributed and parallel computer systems. Dr. Lazowska is a member of the DARPA Information Science and Technology Group, past chair of the Computing Research Association, past chair of the National Science Foundation Computer and Information Science and Engineering advisory committee, and a member of the Technical Advisory Board for Microsoft Research. He served on the CSTB committees that produced *Evolving the High Performance Computing and Communications Initiative to Support the Nation's Information Infrastructure* and *Looking Over the Fence at Networks: A Neighbor's View of Networking Research*. Currently, he serves on the National Research Council committee *Improving Learning with Information Technology*. He is a fellow of the ACM and of the IEEE and is a member of the National Academy of Engineering and a current member of CSTB.

David E. Liddle has a history of conducting and managing computer systems innovation, with an emphasis on interactive systems. At present, after leaving a series of research-management positions, he is a general partner in the firm U.S. Venture Partners (USVP), a leading Silicon Valley venture capital firm that specializes in building companies from an early stage in digital communications/networking, e-commerce, semiconductors, technical software, and e-health. He retired in December 1999 after 8 years as CEO of Interval Research Corporation. During and after his education (he received B.S. and E.E. degrees from the University of Michigan and a Ph.D. in computer science from the University of Toledo, Ohio), Dr. Liddle has spent his professional career developing technologies for interaction and communication between people and computers in activities spanning research, development, management, and entrepreneurship. He spent 10 years at the Xerox Palo Alto Research Center and the Xerox Information Products Group, where he was responsible for the first commercial implementation of the graphical user interface and local area networking. He then founded Metaphor Computer Systems, whose technology was adopted by IBM and the company ultimately acquired by IBM in 1991. In 1992, Dr. Liddle cofounded Interval Research Corporation with Paul Allen. Since 1996, the company formed six new companies and several joint ventures based on the research conducted at Interval. Dr. Liddle is a consulting professor of computer science at Stanford University. He has served as a director at Sybase, Broderbund Software, Metricom, Starwave, and Ticketmaster; he is currently a director with *The New York Times*. He was honored as a distinguished alumnus from the University of Michigan and is a member of the national advisory committee at the College of Engineering of that university. He is also a member of the advisory committee of the school of engineering at Stanford University. He has been elected a senior fellow of the Royal College of Art for his contributions to human-computer interaction. Dr. Liddle has had a number of interactions with national security entities on an advisory basis, providing insights into military mind sets and needs. He is a current member of CSTB.

Tom M. Mitchell has just returned to Carnegie Mellon University (CMU) after a 2-year leave of absence as vice president and chief scientist for WhizBang! Labs. At CMU, he is the Fredkin Professor of Learning and Artificial Intelligence in the School of Computer Science and founding director of CMU's Center for Automated Learning and Discovery. Dr. Mitchell is known for his work in machine learning, data mining, and artificial intelligence. His research ranges from developing software agents that learn to customize to their users, to Web crawlers that learn to extract factual information from Web sites, to computers that mine medical records to learn which future patients are at high mortality risk. He is the author of the widely used textbook *Machine Learning*. Dr. Mitchell is a fellow and president of the American Association for Artificial Intelligence. Prior to joining the faculty of Carnegie Mellon University in 1986, he taught at Rutgers University. He received his B.S. from the Massachusetts Institute of Technology and his M.S. and Ph.D. degrees in electrical engineering from Stanford University. He has had research funded by the Central Intelligence Agency and has consulted with the agency recently about the application of WhizBang! technology to intelligence needs. He is a current member of CSTB.

Donald A. Norman is a user advocate. *Business Week* calls him a cantankerous visionary—cantankerous in his quest for excellence. Dr. Norman is cofounder of the Nielsen Norman Group, an executive consulting firm that helps companies produce human-centered products and services. In this role, he serves on the advisory boards of numerous companies. Dr. Norman is a professor of computer science at Northwestern University and professor emeritus of cognitive science and psychology at the University of California, San Diego. He is a former vice president of the advanced technology group of Apple Computer and was an executive at Hewlett Packard. Dr. Norman is the author of *The Psychology of Everyday Things*, *Things That Make Us Smart*, and, most recently, *The Invisible Computer*, a book that *Business Week* has called the bible of post-PC thinking. He is a current member of CSTB.

Jeannette M. Wing is a professor of computer science at Carnegie Mellon University. Her current focus is on applying automated reasoning tools to specify and verify autonomous and embedded systems for their fault-tolerant, security, and survivability properties. She is the associate dean for academic affairs for the School of Computer Science and the associate department head for the computer science Ph.D. program. She received her S.B. and S.M. degrees in electrical engineering and computer science in 1979 and her Ph.D. in computer science in 1983, all from the Massachusetts Institute of Technology. Dr. Wing's general research interests are in the areas of formal methods, concurrent and distributed systems, and programming languages. She was on the computer science faculty at the University of Southern California (USC) and has worked at Bell Laboratories, USC/Information Sciences Institute, and Xerox Palo Alto Research Center. She has also consulted for Digital Equipment Corporation, the Mellon Institute (Carnegie Mellon Research Institute), System Development Corporation, and the Jet Propulsion Laboratory. She was on the National Science Foundation Scientific Advisory Board and the Defense Advance Research Projects Agency (DARPA) Information Science and Technology (ISAT) Group. She is or has been on the editorial board of seven journals. She is a member of the ACM (fellow), the IEEE (senior member), Sigma Xi, Phi Beta Kappa, Tau Beta Pi, and Eta Kappa

Nu. Professor Wing was elected an ACM fellow in 1998 and is a current member of CSTB.

STAFF

Herbert S. Lin is senior scientist and senior staff officer at the Computer Science and Telecommunications Board, National Research Council (NRC) of the National Academies, where he has been the study director of major projects on public policy and information technology. These studies include a 1996 study on national cryptography policy (*Cryptography's Role in Securing the Information Society*), a 1991 study on the future of computer science (*Computing the Future: A Broader Agenda for Computer Science and Engineering*), a 1999 study of Defense Department systems for command, control, communications, computing, and intelligence (*Realizing the Potential of C4I: Fundamental Challenges*), and a 2000 study on workforce issues in high technology (*Building a Workforce for the Information Economy*). Prior to his NRC service, he was a professional staff member and staff scientist for the House Armed Services Committee (1986-1990), where his portfolio included defense policy and arms control issues. He also has significant expertise in mathematics and science education. He received his doctorate in physics from the Massachusetts Institute of Technology. Avocationally, he is a long-time folk and swing dancer and a poor magician. Apart from his CSTB work, a list of publications in cognitive science, science education, biophysics, arms control, and defense policy is available on request.

Steven Woo is the dissemination and program officer with the Computer Science and Telecommunications Board of the National Research Council. In this capacity, he formulates the dissemination and marketing plan for the study projects and workshops of CSTB. This includes distribution of CSTB reports in government, policy, academia, and private sectors; outreach to promote CSTB to current and potential sponsors; and raising awareness of CSTB's resources and expertise among government and private industry. In addition, he handles the Program Office activities for some of the projects of CSTB. Prior to joining CSTB, Mr. Woo was an Internet and marketing consultant for clients ranging from Fortune 500s to nonprofits. His background includes marketing services for the Los Angeles Dodgers and several years of experience in systems engineering and analysis. Mr. Woo holds a B.S. in engineering from the University of California at Los Angeles and an M.B.A. from Georgetown University.

D.C. Drake has been a senior project assistant with CSTB since September 1999. He is currently working on a project on critical information infrastructure protection and the law and also helped with the project that produced *The Internet Under Crisis Conditions: Learning from September 11*.

He came to Washington in January 1999 after finishing a master's degree in international politics and communications at the University of Kentucky. He earned a B.A. in international relations and German from Rhodes College in 1996. He has worked for the Hanns-Seidl Foundation in Munich, Germany, and in Washington, D.C., for the National Conference of State Legislatures' International Programs Office and for the Majority Staff of the Senate Foreign Relations Committee.

What Is CSTB?

As a part of the National Research Council, the Computer Science and Telecommunications Board (CSTB) was established in 1986 to provide independent advice to the federal government on technical and public policy issues relating to computing and communications. Composed of leaders from industry and academia, CSTB conducts studies of critical national issues and makes recommendations to government, industry, and academic researchers. CSTB also provides a neutral meeting ground for consideration of complex issues where resolution and action may be premature. It convenes invitational discussions that bring together principals from the public and private sectors, ensuring consideration of all perspectives. The majority of CSTB's work is requested by federal agencies and Congress, consistent with its National Academies context.

A pioneer in framing and analyzing Internet policy issues, CSTB is unique in its comprehensive scope and effective, interdisciplinary appraisal of technical, economic, social, and policy issues. Beginning with early work in computer and communications security, cyber-assurance and information systems trustworthiness have been a cross-cutting theme in CSTB's work. CSTB has produced several reports regarded as classics in the field, and it continues to address these topics as they grow in importance.

To do its work, CSTB draws on some of the best minds in the country, inviting experts to participate in its projects as a public service. Studies are conducted by balanced committees without direct financial interests in the topics they are addressing. Those committees meet, confer electronically, and build analyses through their deliberations. Additional expertise from around the country is tapped in a rigorous process of review and critique, further enhancing the quality of CSTB reports. By engaging groups of principals, CSTB obtains the facts and insights critical to assessing key issues.

The mission of CSTB is to:

- *Respond to requests* from the government, nonprofit organizations, and private industry for advice on computer and telecommunications issues and from the government for advice on computer and telecommunications systems planning, utilization, and modernization;
- *Monitor and promote the health of the fields* of computer science and telecommunications, with attention to issues of human resources, information infrastructure, and societal impacts;
- *Initiate and conduct studies* involving computer science, computer technology, and telecommunications as critical resources; and
- *Foster interaction* among the disciplines underlying computing and telecommunications technologies and other fields, at large and within the National Academies.

As of November 2002, current CSTB activities with a cybersecurity component address privacy in the information age, critical information infrastructure protection, authentication technologies and their privacy implications, geospatial information systems, cybersecurity research, and building certifiable dependable systems. Additional studies examine the fundamentals of computer science, information technology and creativity, computing and biology, Internet navigation and the Domain Name System, telecommunications research and development, wireless communications and spectrum management, and digital archiving and preservation. Explorations are under way in the areas of the insider threat, dependable and safe software systems, wireless communications and spectrum management, digital archiving and preservation, open source software, digital democracy, the "digital divide," manageable systems, information technology and journalism, and women in computer science.

More information about CSTB can be obtained online at <http://www.cstb.org>.